Managing Climate Change Business Risks and Consequences

Global Sustainability Through Business

Series Editors:

James A. F. Stoner, *Professor of Management Systems and Chair holder, James A. F. Stoner Chair in Global Sustainability at the Graduate School of Business, Fordham University, New York*

Charles Wankel, *Associate Professor of Management at St. John's University, New York*

The global environmental crisis has been called "the greatest challenge ever faced by our species." In turn, the required changes in business' raison d'être and impacts on the world will be the most profound change in the concept and practice of business ever accomplished. The transformation of business practice must capture and increase for-profit businesses' positive contributions while decreasing and eventually eliminating the harmful effects businesses have on the physical, economic, cultural, social, and political environments of the world. The volumes in *Global Sustainability Through Business* will address the transformations that are occurring and need to occur on a global basis in population, consumption, and production required to achieve a sustainable world—with particular attention to the ways business institutions and business leaders can continue to make their contributions to the world while moving from being "a part of the problem to being a part of the solution." The actions businesses can take start with recognition of business' impacts on the physical, economic, cultural, social, and political environments of the world and include (1) developing business models grounded in sustainability, (2) energizing and motivating businesses and their stakeholders, (3) proactively transforming the environment of business by business, and (4) contributing to reconceptualizing the current dominant paradigms of economic, social, political, and spiritual life. The series will consist of volumes designed to offer the latest cutting-edge research and knowledge about how to forge the future of business organizations in a sustainable world.

Global Sustainability as a Business Imperative
Edited by James A. F. Stoner and Charles Wankel

Managing Climate Change Business Risks and Consequences: Leadership for Global Sustainability
Edited by James A. F. Stoner and Charles Wankel

Managing Climate Change Business Risks and Consequences: Leadership for Global Sustainability

Edited by James A. F. Stoner and Charles Wankel in collaboration with Neil Washington, Matt Marovich, and Kyle Miller

MANAGING CLIMATE CHANGE BUSINESS RISKS AND CONSEQUENCES

First published in 2012 by
PALGRAVE MACMILLAN®
in the United States – a division of
St. Martin's Press LLC, 175 Fifth Avenue, New York, NY 10010.

Where this book is distributed in the UK, Europe and the rest of the world, this is by Palgrave Macmillan, a division of Macmillan Publishers Limited, registered in England, company number 785998, of Houndmills, Basingstoke, Hampshire RG21 6XS.

Palgrave Macmillan is the global academic imprint of the above companies and has companies and representatives throughout the world.

Palgrave® and Macmillan® are registered trademarks in the United States, the United Kingdom, Europe and other countries.

ISBN 978–0–230–11583–5

Library of Congress Cataloging-in-Publication Data

Managing climate change business risks and consequences : leadership for global sustainability / edited by James A.F. Stoner and Charles Wankel.
p. cm.
ISBN 978–0–230–11583–5 (hardback)
1. Business--Environmental aspects. 2. Climatic changes—Risk management. 3. Climatic changes—Economic aspects. 4. Global warming—Economic aspects. 5. Sustainable development. I. Stoner, James Arthur Finch, 1935– II. Wankel, Charles.
HD30.255.M335 2012
658.4'08—dc23 2011035084

A catalogue record of the book is available from the British Library.

Design by MPS Limited, A Macmillan Company

First edition: March 2012

10 9 8 7 6 5 4 3 2 1

Printed in the United States of America.

Contents

Notes on the Contributors

Elizabeth Lokey Aldrich is the chair of the Sustainability Studies Department at White Mountain School. She received her PhD from the University of Colorado in Boulder, where she was an NSF fellow and worked at the National Renewable Energy Laboratory. After she graduated she worked at Camco in emissions trading markets for two years. Currently she is an assistant professor at Boise State University in the Master of Public Policy Program, where she teaches and does research for the Energy Policy Institute. Her articles on renewable energy markets, carbon capture and sequestration, and barriers to the use of offsets in carbon markets are found in *Elsevier's Journal, Renewable Energy, Electricity Journal, Energy Policy*, and *International Journal of Hydrogen*, and her book entitled *Renewable Energy Project Development under the Clean Development Mechanism: A Guide to Latin America* (2009). E mail: lokey@colorado.edu.

Lydia Bals wrote her doctoral thesis at European Business School (EBS), Germany. Research visits took her to University of Pennsylvania, Philadelphia, and Columbia University, New York. Currently, she is project manager at the internal consultancy of the Bayer Group, Bayer Business Consulting. There she also works in procurement and sustainability topics. She is also a visiting scholar at the Department of Strategic Management and Globalization at Copenhagen Business School. Her current research focuses on sustainable sourcing, globalization of R&D (offshore), outsourcing of services, and "born global firms." Her work has been published in such journals as *Journal of International Management, Journal of Supply Chain Management, International Journal of Production Economics, Journal of Purchasing & Supply Management*, and *Industrial Marketing Management.* Email: lydia.bals@web.de.

Ann Brockett is the Americas director of Climate Change and Sustainability Assurance Services at Ernst & Young LLP. As a practice leader and subject-matter professional with more than 17 years of experience, Ann has serviced public and private clients across a range of industries on climate change, carbon management, and sustainability. With a wealth of experience in identification and development of alternatives in emerging risk areas, Ann has provided greenhouse gas verification services and reporting under the Alberta Legislation since its inception. She has led assurance services teams to provide large emitters with effective verification standards for their baseline inventory submissions to the Alberta government. Ann has also provided sustainability assurance services leadership in the tobacco manufacturing industry, focusing on climate-change-related indicators. She has been involved in sustainability legislation in Alberta, for nearly two decades, and her extensive knowledge of the implementation challenges of regulatory programs led to her participation in a "lessons learned" session to discuss the results from the first year of mandated reductions and assurance requirements under the Alberta greenhouse gas reduction program. In recent years, Ann has provided advisement regarding the impacts on Canadian industries from new carbon regulatory and market schemes. She has also participated in a joint climate change information session with the Washington Council, with a focus on the current North American regulatory regimes and future trends for oil and gas executives. E mail: ann.m.brockett@ca.ey.com.

Lára Jóhannsdóttir holds a BS in Business Administration and an MBA in Global Management (with honors) from the Thunderbird School of Global Management. Currently she is a doctoral candidate at the University of Iceland, affiliated member of NORD-STAR, the Nordic Center of Excellence for Strategic Adaptation Research, board member of a pension fund, and a part-time university lecturer. Her research focus is Nordic nonlife insurance companies and environmental issues, particularly climate change. She is an independent consultant and a business owner of Alvör, a start-up company focusing on environmental strategies of nonpolluting companies. She has 14 years of working experience as a specialist and manager for Icelandic insurance companies. Among her publications are "Green office practices" in Ted Hart's *Nonprofit Guide to Going Green* (2009), "Woe awaits insurers! The Nordic insurance industry and climate change" in Ingjaldur Hannibalsson's *Rannsóknir í félagsvísindum X. Reykjavík: Félagsvísindastofnun Háskóla Íslands* (2009), and *Environmental literacy of business students* (2009). E mail: larajoh@gmail.com.

Aled Jones is the director of the Global Sustainability Institute (GSI) at Anglia Ruskin University. The GSI is a University-wide body, which spans a broad portfolio of areas including built environment, technology, ecosystems, business, education, and health. Dr. Jones chairs a working group on climate finance within the Capital Markets Climate Initiative on behalf of Greg Barker, the minister for Climate Change in the UK Department for Energy and Climate Change and sits on the UNEP FI insurance sector working group. He was a founding member of the ClimateWise insurance principles and was facilitator for the P8 Pensions Group, 12 of the world's largest public pension funds working collaboratively to address the problem of climate change. E mail: aled.jones@anglia.ac.uk.

Aimée A. Kane holds a PhD and an MS in Organizational Behavior and Theory from the Tepper School of Business at Carnegie Mellon University and a BA in Spanish, magna cum laude, and a certificate in markets and managements studies from Duke University, where she was elected to Phi Beta Kappa. Currently she is an assistant professor of Management at the Palumbo-Donahue School of Business at Duquesne University. Her research, which focuses on knowledge transfer and creation, social identity, newcomers, and sustainability, has appeared in several top publications, including *Organization Science, Organizational Behavior and Human Decision Processes, Best Paper Proceedings of Academy of Management*, as well as in edited books. She is a member of the editorial review board of *Organization Science*. Before coming to Duquesne, she was an assistant professor of Management and Organizations at New York University Stern School of Business. E mail: kanea@duq.edu.

Cassandra Koerner received her BS in Natural Resource Conservation from the University of Montana. Currently she is a graduate student at Boise State University and is employed by the Energy Policy Institute. Prior to coming to Boise she worked in the public forest sector in Montana and Washington. Cassandra is interested in global climate change, energy and natural resource policy, economics, and social responses to scientific evidence and their related policy changes. E mail: cassiekoerner@boisestate.edu.

Nicole Kranz holds a PhD in Political Science from Freie Universität Berlin as well as Masters degrees in Environmental Science and Management from German and American universities. She currently works as a research associate with the Center of Transnational Relations, Foreign and Security Policy at the Institute for Political Sciences of Freie Universität Berlin and also as a policy advisor with German International Cooperation (GIZ) in the area of international water policy and infrastructure with a focus on UN processes for improving access to water and sanitation. Her research focuses on governance in areas of limited statehood with a specific emphasis on the role of private actors and state capacity in sustainable development. Empirically, a significant part of her past and current work revolves around issues of water governance and adaptation to climate change, with a specific focus on the situation in Africa. E mail: nicole.kranz@fu-berlin.de.

Candace A. Martinez received her PhD in Strategic Management from the University of Illinois at Urbana-Champaign. Currently she is an assistant professor of International Business (IB) at Saint Louis University. Dr. Martinez's research focuses on the influence of formal and informal norms in institutional environments across countries. Her current research examines the role of institutions in encouraging the development of sustainability objectives and in governing the informal waste collection/recycling sector in Latin America. She has recently published a book chapter in *Global Sustainability as a Business Imperative* (ed. James A. F. Stoner and Charles Wankel) on informal waste-pickers in Latin America and an article, with G. Allard, on the influence of social policy on foreign direct investment in developing countries in *The Journal of Business in Developing Nations*, and a case study on a social entrepreneurship venture in Guatemala City in *Cases and Exercises in Organization Development and Change* (ed. Donald L. Anderson). E mail: cjmartnz@gmail.com.

Travis L. McLing is a doctoral candidate at the University of Idaho, Moscow. He began working at the Idaho National Laboratory (INL) in June 1992 as a geochemist and over the years has worked on a variety of projects associated with water/rock interactions in groundwater systems. His current research duties have a diversity of objectives but deal largely with carbon dioxide sequestration, and its associate technical issues. Travis is currently the Technical Lead for Carbon Sequestration at the INL and is a sitting member of the Idaho Governors Carbon Advisory Committee as well as the Chairman of the Idaho Carbon Issues Task Force. He has been conducting research and public outreach in the field of carbon issues and carbon sequestration since 2002. His primary interest in this field is the study of the geochemical mineralization reactions controlling the fate and transport of carbon dioxide in subsurface environments. In addition to his research activities Travis' responsibilities also require him to interact with industry, education, and media entities that have interest in carbon dioxide mitigation. E mail: travis.mcling@inl.gov.

Joseph C. Perkowski earned his PhD from MIT in Civil Engineering/Environmental Systems Management. Currently he is manager of Energy Initiatives at the Idaho National Laboratory in Idaho Falls. Prior to joining INL, he served on assignment to the National Renewable Energy Laboratory (NREL) as Market Sector manager, responsible for the integration of selected technical activity in renewable energy technology with the commercial marketplace. He worked at the Bechtel Corporation before his NREL assignment as Manager of Advanced Civil Systems Research and Development, and prior to that at United Technologies Corporation (UTC) and at the Oxford Development Group of Edmonton, Alberta. Earlier employment included the position of senior research officer with the Corporate Environmental and Social Affairs Department of Petro-Canada, with responsibilities that included developing internal business policy papers and guidelines on environmental impact assessment techniques. At INL Perkowski currently works primarily with private sector clients (such as large North American utility firms) on a variety of new technology development issues, including carbon sequestration liability assessments. E mail: Joseph.Perkowski@inl.gov.

Efrain Quiros, III has a Master of Human Relations from the University of Oklahoma, a Master of Science in International Business Economics (with Distinction) from Anglia Ruskin University, and a Bachelor of Science in Business Administration from California State Polytechnic University, Pomona. Currently he is an independent researcher and collaborates with the Lord Ashcroft International Business School in Cambridge, UK. He is an associate of the Leadership for Sustainable Development Programme with LEAD International, an associate of the Institute for Environmental Management and Assessment (IEMA), and a member of the Chartered Management Institute (CMI). He has held various managerial roles in business operations, including operational risk management, across a range of commercial sectors. E mail: Efrain.Quiros@gmail.com.

Zabihollah Rezaee received his BS from the Iranian Institute of Advanced Accounting, his MBA from Tarleton State University in Texas, and his PhD from the University of Mississippi. Currently he is the Thompson-Hill Chair of Excellence and professor of Accountancy at the University of Memphis and has served a two-year term on the Standing Advisory Group (SAG) of the Public Company Accounting Oversight Board (PCAOB). Professor Rezaee holds several certifications, including Certified Public Accountant (CPA), Certified Fraud Examiner (CFE), Certified Management Accountant (CMA), Certified Internal Auditor (CIA), Certified Government Financial Manager (CGFM), Certified Sarbanes-Oxley Professional (CSOXP), Certified Corporate Governance Professional (CGOVP), and Certified Governance Risk Compliance Professional (CGRCP). He has been finalist for SOX Institute's SOX MVP 2007, 2009, and 2010 award. Professor Rezaee has published over 180 articles in a variety of accounting and business journals and made more than 200 presentations at national and international conferences. He has also published seven books: *Financial*

Institutions, Valuations, Mergers, and Acquisitions: The Fair Value Approach; *Financial Statement Fraud: Prevention and Detection*; *U.S. Master Auditing Guide* 3rd edition; *Audit Committee Oversight Effectiveness Post-Sarbanes-Oxley Act*; *Corporate Governance Post-Sarbanes-Oxley: Regulations, Requirements, and Integrated Processes*; *Corporate Governance and Business Ethics and Financial Services Firms: Governance, regulations, Valuations, Mergers and Acquisitions.* His latest book, *Business Sustainability and Accountability Reporting*, is scheduled to be published in 2011. E mail: zrezaee@memphis.edu.

Jonathan Smith has a PhD from Hull University in the UK, an MA in Human Resource Management, and a BEng in Engineering. He is a senior lecturer at the Lord Ashcroft International Business School where he has led a number of leadership and HRM programs. He designs and facilitates innovative Master's level courses in leadership, strategy, organizational change, HRM, and research methods. Jon coaches and supports leaders and HR professionals in the research, design, and implementation of best practice initiatives in organizations. His current research, consulting, and development interests are focused on organizational, team, and individual transformation through spiritual leadership; development and training within the police; and sustainable development for businesses. Jon has experience in a variety of managerial and training roles in a number of public and private sector organizations. Prior to working at the University, Jon was a Director of Studies at the UK National Police Training and Development Authority, working to shape the training agenda to achieve world-class performance within police training. He is a chartered fellow of the Chartered Institute for Personnel and Development (CIPD) and a fellow of the HE Academy. E mail: Jonathan.Smith@anglia.ac.uk.

Alison L. Steele holds a BA in Physics from Drew University and is a recent graduate of the MBA-Sustainability program at Duquesne University's Donahue Graduate School of Business. Her interest in sustainability began during her undergraduate studies when she explored the impacts of deforestation on regional communities as part of a month-long ecotourism seminar in Ecuador. As a Sustainability research fellow at Duquesne, she has assisted in developing a microenterprise plan for waste reduction and job creation in Ecuador and has investigated the connection between components of organizational behavior and the successful adoption of sustainability initiatives. As an MBA student, she has worked on two consulting projects at local organizations: developing a sustainable product review framework for a healthcare technology company and organizing a greenhouse gas displacement database for a nuclear energy provider. She hopes to leverage her expertise in physics and sustainability to improve living conditions in developing nations. E mail: alisonlsteele@gmail.com.

James A. F. Stoner earned his BS in Engineering Science at Antioch College in Yellow Springs, Ohio, and his SM and PhD in Industrial Management at MIT, in Cambridge, MA. Currently he is professor of Management Systems at Fordham University's Graduate School of Business and chair holder of the James A. F. Stoner Chair in Global Sustainability—a chair endowed in Jim's name by one of Jim's students (Brent Martini) and his father (Bob Martini) in acknowledgment of Jim's teaching and research at Fordham and his contributions to their and their company's work. Jim was chief of a volunteer fire department, and was the Government of Tanganyika's first project development officer as an MIT Fellow in Africa. He has published articles in such journals and periodicals as *Academy of Management Review, Harvard Business Review, Journal of Development Studies, Personnel Psychology*, and *Journal of Experimental Social Psychology*, and has authored, coauthored, and coedited 20 books. Jim's current projects and interests focus on finding ways to move toward a sustainable world. He has taught managers, executives, MBA and undergraduate students in at least a dozen countries including Brazil, Ethiopia, Iran, Ireland, Japan, and Russia. He has won teaching awards at Fordham and has consulted with a broad range of companies, including Bell Labs, Richardson-Merrill, BergenBrunswick, and Arthur D. Little, Inc. E mail: stoner@fordham.edu.

Sujit Sur received his PhD in Strategic Management from Concordia University and is currently an assistant professor at Dalhousie University, Halifax, Canada. His research focuses on the implication of corporate ownership, international governance, and corporate reputation on firm performance, including sustainability-related initiatives. Sujit's research is published in *Journal of Management and Governance*, as well as in the proceedings of *European Group for Organizational Studies, Administrative Sciences Association of Canada*, and *Eastern Academy of Management.* Sujit also has coauthored book chapters in *Contemporary Management: Collaborating in a Networked World* (forthcoming) and *Encyclopédie de la stratégie* (forthcoming). He has also presented his work at the *Academy of Management, Strategic Management Society* and *Decision Sciences Institute* annual meetings. Additionally, Sujit has organized symposiums on the strategic imperative of sustainability at the *Academy of Management* as well as the *Administrative Sciences Association of Canada's* annual meetings. E mail: sujitsur@dal.ca.

James Wallace holds an MSc in Environmental Technology from Imperial College specializing in Pollution Management and Control. Currently he is group head of Corporate Responsibility at RSA Insurance Group plc. James is responsible for the integration of environmental and social factors across operations in 34 countries and underwriting business in over 130 countries. Areas of focus include investments, procurement, underwriting, environmental management, community investment, nonfinancial reporting, communications, and more. He is responsible for developing one of the largest international corporate partnerships with the World Wildlife Fund focusing on the insurance risks of environmental change. He represents RSA on the Financial Services Corporate Responsibility Group focusing on responsible procurement, is a member of the insurance working group of UNEP-FI and mentors small businesses starting Corporate Responsibility programs in London. He is an associate of the Institute of Environmental Management and Assessment as well as being a registered BREEAM assessor for offices. Prior to joining RSA, James worked as a Corporate Responsibility and Environmental consultant working for a range of FTSE 100 corporations and was a graduate with HSBC. Email: James.Wallace@gcc.rsagroup.com.

Charles Wankel, associate professor of Management at St. John's University, New York, earned his doctorate from New York University. Charles currently is on the Rotterdam School of Management Dissertation Committee and is Honorary vice rector of the Poznan University of Business. Recent books include *Global Sustainability as a Business Imperative*, *Innovative Approaches to Global Sustainability*, *Global Sustainability Initiatives*, *Alleviating Poverty through Business Strategy*, *Innovative Approaches to Reducing Global Poverty*, and the *Encyclopedia of Business in Today's World*, which was recently selected by the American Library Association as Outstanding Business Reference Source of the Year. He is the founder and director of several scholarly virtual communities including the Managing-for-Sustainability interdisciplinary discussion forum (1,100 participants), which you are invited to join. He has been a visiting professor in Lithuania at the Kaunas University of Technology (Fulbright Fellowship) and the University of Vilnius (United Nations Development Program and Soros Foundation funding). Email: wankelc@stjohns.edu.

Figures

Tables

PART I

Climate Change and Risk Management

CHAPTER 1

Framing the Inquiry into Risk Management and Climate Change

James A. F. Stoner and Charles Wankel

Smart companies now recognize that tolerating wasteful energy use and higher carbon emissions is a high-risk strategy. Geopolitical volatility, the unpredictability of energy supplies, price increases, threats to business from extreme weather events, and the risks of liability claims for failing to manage carbon output all make carbon reduction a good business strategy. The FTSE Index, the British equivalent of the Dow Jones Industrial Average, put it succinctly. "The impact of climate change is likely to have an increasing influence on the economic value of companies, both directly, and through new regulatory frameworks. Investors, governments and society in general expect companies to identify and reduce their climate change risk and impacts, and also identify and develop related business opportunities."

Lovins and Cohen (2011:8)

The authors of the chapters in this book all address the question of what climate change *does* mean for business and what business practices and strategy *can* mean for climate change. The jury is now clearly in on climate change. It no longer can be seen as something that might happen to our grandchildren; it is something that is impacting all of us now.

This volume uses the term climate change in its title and in the various chapters it contains. An equally appropriate term might be "global sustainability," or better yet "global unsustainability." Although the most pressing of all presently apparent aspects of global unsustainability is climate change—driven by the continually rising level of CO_2, a level that was described in the 2010 Second Global Forum for Responsible Management Education (PRME) at Fordham University as the "390 pound gorilla" that sits in the back of our classrooms ignored by faculty and students collaborating in a conspiracy of silence as they (we) pursue our commitments to succeeding at "business as usual."

That "gorilla" weighed "only" 392 pounds in June 2010 but by the time this book is first available to you, it is expected to weigh about 395 pounds, as CO_2 levels continue to rise. Although the rise of atmospheric CO_2 levels above the estimated 350 ppm upper limit of a "safe" level for the earth's bio-systems is apparently the gravest current threat to our species' comfort and perhaps existence, the issue of global unsustainability is much broader than the level of CO_2 in the atmosphere. The authors in this volume recognize that climate change is one aspect of the integrated, holistic "wicked problem" (Churchman, 1967) of global unsustainability. That wicked problem includes global warming, deforestation, biodiversity loss, species' extinction, ocean acidification, aquifer contamination, coral reef destruction, soil erosion and soil poisoning from chemical-based fertilizers, pesticides and monoculture, glacier melting, etcetera.

Although the problems of global unsustainability are myriad, climate change deserves our particular attention for two reasons. First, we can do something about it. In our Anthropocene era, our actions have been changing the climate and we can do things to "unchange it" or at least to stop changing it for the worse to some extent. Second, what is "special" about climate change is that it has an influence on many other aspects of global unsustainability. Changes in weather patterns are effecting food production, which is adding to global poverty. Attempts to reduce our dependency on fossil fuels to limit carbon emissions have led to ineffective and inefficient processes for converting corn crops to ethanol and doing so has very likely also added to food shortages and higher prices for food in areas of the world that can least afford such price increases. And on and on and on. So, dealing with the risks and consequences of climate change is what we should be doing. And perhaps it is the most important of all things we should do—arguably the "only game big enough for us to play" (Stoner and Wankel, 2009:3).

In addressing the ways companies can manage the risks and consequences of climate change, the authors of the chapters that follow are correct in focusing our attention on the most pressing problem of the twenty-first century—how we can find ways, particularly through wise, inspired, caring, and even selfish business leadership to stop that gorilla from growing ever fatter and eventually slim it down to a svelte 350 pounds or even a bit less.

The volume's first section, "Climate Change and Risk Management," contains two chapters that provide the framework for looking at the business risks and consequences of climate change within which the next nine chapters make their contributions. This chapter, Chapter 1, introduces the broad topic of managing the business risks and consequences of climate change and provides a chapter-by-chapter overview of key themes in the chapters that follow. The overview of each chapter highlights key points and important conclusions to be found in each chapter. Chapter 2, by Sujit Sur, provides a framework for organizing business leaders' and our understanding and perceptions of risk.

The six chapters in the second section of the book focus on the key institutions working to manage the risks of climate change. In Chapter 3 Lára Jóhannsdóttir, James Wallace, and Aled Jones focus on the risk management industry and discuss the primary insurance industry's role in managing climate change risks and opportunities. Chapter 4, by Jonathan Smith and Efrain Quiros, III takes the perspective of corporate managers and focuses on how corporations can capitalize on the economic benefits of working toward climate change mitigation and global sustainability and, in doing so, achieve competitive advantages through effective climate change risk management. Chapter 5, by Lydia Bals, continues the focus on corporate managers' responsibilities to be risk managers and opportunity finders by exploring the risks and opportunities arising from climate change's impact on the procurement process of companies, and all organizations. In Chapter 6 Nicole Kranz addresses the adaptation strategies businesses can use to respond to climate change and how they can work with local stakeholders. Her chapter reports ways some firms in the South African mining industry have been working with the national government and local governing bodies to deal with the impacts of climate change on water supply. Chapter 7 continues the focus on how corporate climate risk management decisions are made by shifting attention to the context in which those decisions are influenced by concern for external stakeholder pressures. In that chapter, Ann Brockett and Zabihollah Rezaee emphasize the growing importance of the sustainability reports companies and other organizations prepare and describe one of the new ways such reporting is being organized. The final chapter of this section of the volume describes how private companies, an NGO, and local informal entrepreneurs have been successful in reducing the risk of climate change in Latin America. In that chapter Candace A. Martinez discusses the ways ineffective handling of waste contributes to climate change risk and how informal waste recyclers organized themselves, with the aid of the private sector and an NGO, to improve the collection and recycling of waste products in Bogotá, Colombia. Her chapter calls attention to one of the many subtle and easily overlooked ways poverty alleviation and the reduction of social injustice can contribute to mitigating climate change risk.

The third and final section of the book deals with emerging climate risk management issues and possible ways climate change and global sustainability risk management might evolve. In Chapter 9 Elizabeth Lokey Aldrich, Cassandra Koerner, Joseph Perkowski, and Travis L. McLing describe how the world is grappling with a particularly complex possibility for reducing the growing levels of CO_2 in the atmosphere—carbon sequestration. They provide an up-to-date report on the liability concerns and alternatives involved in attempts to manage the risks and opportunities of this emergent technology. In Chapter 10 Aimée A. Kane and Alison L. Steele address the very important issue of making something happen. They identify an important part of the pathway companies need to follow as they mobilize themselves to take actions to deal with climate change risks and opportunities. Finally, in Chapter 11, the editors summarize some of the key aspects of the "wicked" climate change problems and discuss opportunities for leadership in managing the risks that problem raises for all of us as we work to move toward a more sustainable world.

Part I: Climate Change and Risk Management

In the other chapter in Part I, Chapter 2, "Asking the Right Questions: Basic Concepts of Risk Management for Developing Adaptation and Mitigation Strategies," Sujit Sur provides a framework for organizing business leaders' and our understanding and perceptions of risk. Building on that framing, he shows how the nuances of those understandings and perceptions impact the decision-making processes businesses must use to develop mitigation and adaptation strategies as they work to manage climate change business risks. The main theme of the chapter is that unless corporations adopt the appropriate mindset for understanding the nature of climate change risk, their attempts at mitigating climate change concerns will be ineffective. The discussion distinguishes the three classical ways of framing our understanding of climate risk, or any other type of risk situation: pure "risk," "ambiguity," and "uncertainty." The chapter's key insight for business leaders is that corporate decision-making is predominantly geared toward "risk" mitigation while the conditions regarding climate change are "uncertain" or "ambiguous." Thus corporations need to change their mind-set to an uncertainty/ambiguity framing as they develop their mitigation strategies. Doing so will enable them to use entrepreneurial, creative, and innovation-based approaches in developing their climate change mitigation and adaptation strategies.

The first section of the chapter builds on Frank Knight's work distinguishing between risk, ambiguity, and uncertainty based on the distinction between knowledge of possible outcomes and the probabilities of the outcomes (Knight, 1921). A third component is added to the basic 3×2 model in Knight's work to integrate the consequence of outcomes, resulting in a 3×3 table. The 3×3 table contains probabilities, outcomes, and consequences of the outcomes. The table is used to frame the discussion of how the interplay of these components leads to the complexity we are experiencing in our decision-making about ways to deal with climate change uncertainties.

The chapter then provides brief introductions to five behavioral perspectives on individual/organizational decision-making under complex/uncertain conditions: risk society, edgework, systems theory, cultural theory, and governmentality. It shows how the dynamics arising from these perspectives can contribute to suboptimal decisions and to changing our perception of time.

These concepts are integrated into a framework that can help us understand our present differing perspectives on climate change (why we lack consensus on realities and desirable actions and why the rhetoric is becoming increasingly polarized), and can provide guidelines for choosing appropriate mitigation strategies. The concepts are then used to develop a typology of business risk mitigation strategies to reduce the impacts of climate change.

The chapter highlights the need to shift the climate change conversation toward opportunities to create a better world and away from the present rhetoric on the "doom and gloom" risks of climate change. Such a shift would focus on what can be, instead of the present focus on what is not. To paraphrase Alex Steffen (2006), "We need to provide a third choice between being poor and frugal and being rich and wasteful. And we need to do it very, very quickly."

Part II: Key Players Managing the Risks of Climate Change

The overarching message of Chapter 3, "The Primary Insurance Industry's Role in Managing Climate Change Risks and Opportunities," by Lara Jóhannsdóttir, James Wallace, and Aled Jones is that climate change is creating huge risks, that the insurance sector is not being sufficiently proactive in confronting those risks, and that climate change risks have implications for far more of their business than companies in the insurance sector realize. However, that overarching message does include the fact that there are also great opportunities to tackling these risks, and if done well, such proactive responses will lay the basis for a very healthy insurance sector for the future. The chapter addresses three questions: (1) What does climate change risk mean for insurers? (2) How can climate risk be dealt with and/or reduced? And (3) What are the climate change opportunities?

In addressing the first question, what climate change risk means for insurers, they describe how climate change presents a number of risks to insurers, the economy, and society including physical, market, policy, and security as well as fiduciary risk on the investments of insurance companies. Some of these risks are more immediate than others but all will need careful management over time. These risks impact all aspects of the insurance sector from general to life insurance as well as professional indemnity and security-related insurance. The chapter explores some of the emergent risks, the impacts they might have on various insurance business lines, and how they may evolve with time.

For the second question, "How can climate risk be dealt with and/or reduced?," they note that the insurance sector has been identified as one of the sectors highly exposed to climate change because of socioeconomic changes, such as growing world population and urbanization, and more frequent and severe weather events. Climate risk endangers different lines of insurance business, such as life and health, property and casualty, investments, liability and business interruption. In response to this question, they explore the ways insurers can reduce or cope with climate change risk.

In addressing the third question, dealing with climate change opportunities, they report that for proactive insurers climate change offers significant opportunities for creative solutions, as opposed to being just a threat to their operating conditions. The insurance sector has been identified as being a leading sector in this respect. Insurance response to climate change has been tracked since 1999 by Ceres, and studies show that the number and depth of initiatives in this industry are rising. The chapter closes with an exploration of various ways insurers can take advantage of the opportunities created by climate change.

In Chapter 4, "Competitive Advantages and Risk Management: Capitalizing on the Economic Benefits of Sustainability," Jonathan Smith and Efrain Quiros, III emphasize how climate change and environmental issues raise pressing global concerns that make working toward a sustainable world a necessity for all stakeholders and shareholders alike. They note that environmental issues are relevant to companies and managers across all industries and sectors, and that businesses must increasingly consider how they utilize resources effectively to minimize impacts on the climate and the environment in general. A focus on a wider definition of sustainability concerns can benefit the fundamental economic interests of business. Rather than seeing climate change and global unsustainability as only a threat, companies can respond to the business opportunities for mitigation and adaptation and, in doing so, can improve business models, activities, and approaches.

They describe how a new paradigm in the global business environment has evolved, making commitments to reducing the risks of climate change and aligning plans and actions with the need for global sustainability a business imperative in ways that cannot be ignored by management. Whereas the old view pitted environmental interests against business interests, a new paradigm links these two "interest groups" as interdependent with mutual interests. Environmental concerns are part of the business environment and, thus, cannot be ignored by companies. Through the implementation of various green business strategies, companies can fulfill economic responsibilities with activities that are more environmentally, socially, and economically friendly.

Drawing on recent research on HSBC and Walmart, the authors show how the companies' sustainability commitments are providing improved business approaches and addressing a wide range of risks in the business environment. The companies are actively utilizing sustainability-supportive initiatives for profit maximization and to gain competitive advantages. They are also doing so with varying degrees of engagement and success.

The authors conclude that managers cannot reject "green business" based on misinformed beliefs that sustainability is inherently opposed to business interests. Companies should seek to incorporate sustainability-focused initiatives whenever possible for the sake of their own economic interests, not solely to serve environmental interests. Although managers should actively integrate sustainability into business activities as much as possible, engagement with sustainability is situational and circumstantial, and should vary depending on company activities. Not all companies can be completely carbon-neutral, nor do they have to be. However, the authors argue that because sustainability-focused initiatives can be powerful vehicles for gaining competitive advantages and for capitalizing on economic opportunities, companies and managers are fundamentally obligated to engage in sustainability-driven initiatives.

Chapter 5, "Climate Change Impact on Procurement: Risks and Opportunities," by Lydia Bals turns the focus to the procurement process in companies (and in any organization) and addresses two major themes related to acquiring the resources and supplies needed to conduct productive operations. The first theme is the implications of climate change for the situations faced by and actions of companies acquiring the resources and supplies they need to conduct their businesses. The second concerns the implications climate disruptions are likely to have for what it means to do "sustainable business," with a particular focus on the implications for the companies' sourcing function.

In the chapter, Bals notes that supply markets are becoming progressively more affected by the increasing impacts of climate change. In understanding these impacts and businesses' responses to them, it is useful to distinguish goods markets from services markets. Whereas goods, and physical inputs in general, are subject to logistical/physical flow considerations and natural resource market considerations, services are not. Therefore, she discusses the two markets for inputs to businesses separately and clarifies the differences in the implications of climate change for each.

Turning toward the first theme and focusing on production-oriented businesses, climate changes are already having a considerable impact on natural resource markets, leading to higher market price volatilities. Supply distortions caused by "natural disasters," some of which may be triggered or even caused fully by climate change, are leading companies to cope with insecure supply situations; potentially having severe implications for their own production. Examples of some of these disruptions include floods (e.g., flooded coal mines in Australia recently) and losses of wheat production. Such events lead to subsequent effects on the supply chain (e.g., resulting increases in steel prices and food prices). She notes that these impacts can be especially severe in production-oriented industries that operate with (close to) zero stock policies (such as the automotive industry).

In addressing the second theme, she notes that the procurement function's role in facilitating sustainable business involves—in its simplest interpretation—ensuring supply security to maintain the company's ability to produce its products and services. However, she emphasizes that procurement's role can and should go far beyond that narrow interpretation. Opportunities for a much expanded role for procurement arise in part because the rules for business are changing: suppliers are affected by new restrictions, customers' attitudes change, regulations change, and demand for and supplies of more sustainability-friendly products are growing. Monitoring the supply market implications of these business environment changes is an important task for businesses. However, in a world of climate change, monitoring what is happening in customers and suppliers that are leaders in providing sustainability-friendly products and services (e.g., products and services with the smallest possible resource footprint) and learning from them becomes a vital opportunity for businesses to open up new market segments or completely new markets. The procurement function is ideally placed for conducting that monitoring and for capturing that learning.

The market demands for sustainable products and services need to be analyzed in conjunction with regulatory requirements now in place or in the making. The increasing importance of building "sustainable products" as a source of competitive advantage for a growing market puts a high emphasis on learning from suppliers as sources of innovation. It will also be necessary to collaborate with suppliers to help them cope with climate change to avoid disruptions in the company's supplies and also to ensure that suppliers are following sustainable practices so one's own products and the supply chain they are based on warrant the label "sustainable."

Bals notes that what it means to be a "sustainable business" will probably move beyond a carbon only interpretation and more toward the UN/Brundtland Commission definition (World Commission on Environment and Development, 1987) that includes the ecosystem dimension, as well as the social dimension (which it already does at some companies). Such an evolution of meaning will be a corporate social responsibility (CSR) opportunity for businesses and their suppliers, offering ways to differentiate them from the competition. The challenge is to bring to this opportunity an aligned effort with suppliers, so sustainable practices can be ensured along all supply tiers. Bals believes that seizing this opportunity effectively will require new models of coordination and cooperation, but those models will yield significant advantages when they are developed.

In Chapter 6, "Business and Climate Change Adaptation: Contributions to Climate Change Governance," Nicole Kranz starts by acknowledging that business is commonly considered as the main culprit for climate change, in large part due to its climate-relevant emissions. While this view has led to widespread criticism of business practices, there is growing awareness that a major contribution of the business community is necessary to address current and future climate change challenges. Determining how such a contribution can occur requires an analysis of the role business can play in this regard and of the motivations that drive business behavior.

While academic literature on the role of business in responding to climate change focuses largely on the strategic orientation at the firm level, with a strong emphasis on mitigating climate-relevant emissions, adapting the business organization to cope with impacts of climate change on operations, and creating a more resilient business entity, Kranz takes a different perspective. In emphasizing that environmental change affects business and the surrounding communities alike, she focuses attention on the interaction between these two groups. To explore the roles that businesses can play in managing the risks of climate change and their motivations for doing so, Kranz describes how private companies in two South African mining districts have been working with local authorities and the national government to deal with company and community needs for secure and clean water supplies.

Her discussion of coal mining companies' experiences in the Mpumalanga Highveld and platinum mining companies' experiences in the country's Bushveld complex covers firm-internal adaptation measures, as well as measures that originate from the companies but have strong external implications for the communities. These involvements with the surrounding communities are targeted at improving overall adaptive capacity and resilience for both the companies and the communities. In this regard, she finds that company measures transcend the immediate sphere of influence of the firms and contribute to overall climate change governance.

Her chapter also investigates governance structures that motivate these and other businesses to embark on adaptive strategies. Specific focus is placed on the interaction between public and private actors and this focus suggests useful insights for sustainable development governance.

In Chapter 7, "Sustainability Reporting's Role in Managing Climate Change Risks and Opportunities," Ann Brockett and Zabihollah Rezaee describe "business sustainability" as a process that enables organizations to establish strategies that will improve multiple bottomline performance in five areas: economic, governance, social, ethics, and environment (EGSEE). They emphasize the growing importance of sustainability reporting for companies and other organizations and describe one of the new ways such reporting is being organized. The chapter is intended to provide managers with ideas for rethinking their organizations' overall business strategies by integrating sustainability development into their core values, actions, risk management, corporate governance, and corporate reporting. They note that in the post 2007–2008 global financial crisis era, financial information per se no longer satisfies the needs and demands for financial and nonfinancial information on key performance indicators (KPIs) for EGSEE dimensions of business sustainability.

No business can grow without taking proper risks in managing its operations. The type and level of risk a business is willing to assume is commonly known as its risk appetite or risk tolerance. In this chapter, they discuss three risks (operations, compliance, and reputation) relevant to sustainability reporting and climate change concerns. In their discussion of sustainability reporting they include consideration of accounting, assurance, and disclosure of sustainability strategies, risk assessment, and activities.

They note the importance of climate change and that business sustainability has emerged as the central theme of the twenty-first century, growing out of its roots in corporate governance. It is ultimately the responsibility of corporate boards to focus on business sustainability for creating enduring value for shareholders and for protecting the interests of other stakeholders, including creditors, employers, suppliers, government, and society at large. They see business sustainability and corporate accountability as being "all about" adding value in all areas of EGSEE matters and events. As part of the growing awareness of business' broad and encompassing responsibilities, many public companies are now voluntarily managing, measuring, recognizing, and disclosing their commitments, events, and transactions relevant to climate change and EGSEE. Business sustainability is seen not only as ensuring long-term profitability and competitive advantage for the organization but also as maintaining the well-being of the society, the planet, and its people.

In Chapter 8, "Climate Change Risk and Informal Recycling: An NGO and Private Sector Partnership in Bogotá," Candace A. Martinez describes the crucial and often unrecognized role that developing world informal recyclers play in reducing the risks associated with climate change. She describes a pilot program launched by four multinational corporations, a nongovernmental organization, and informal recyclers in a Latin American city. The program serves as an example of how innovative private companies, NGOs, and local community members can create partnerships that reduce environmental, economic, and social risks and in doing so contribute to mitigating the risks of climate change.

Martinez notes that many such programs and similar initiatives are needed because the consequences of not reaching global sustainability goals are serious across the developing world but are perhaps direst for Latin America. The region produces approximately 6 percent of the world's global greenhouse emissions (Tuck, 2009), yet if worldwide emissions are not brought under control, it is forecast that Latin America will suffer a disproportionate number of negative impacts. According to a December 2009 report from the Economic Commission for Latin America and the Caribbean (ECLAC), if the planet were to continue at its current rate of greenhouse gas emissions, Latin America could experience melting glaciers in the Andes mountains wreaking havoc on farmlands, rising sea levels causing mangrove forests to disappear, degraded lands claiming from 22 percent to 62 percent of Bolivia, Chile, Ecuador, Paraguay, and Peru, and the biodiversity of Colombia and Brazil becoming severely depleted. The report suggests the importance of partnerships like the one described in this chapter by observing that improving solid waste management in developing countries yields one of the most effective weapons with which to prevent an environmental Armageddon (Estrada, 2009).

Martinez discusses how the types of partnerships she describes can reduce the risk of global warming and also contribute to reducing the social risks of global unsustainability, such as poverty and the disenfranchisement of low-income populations. She shows how businesses, by leveraging the expertise of informal recyclers, can also contribute to practical sustainability (environmental, economic, and social) gains. Her goals in the chapter go beyond describing the contributions such approaches can make to mitigating climate change risks. She wants to provide readers with an introduction to the world of informal recyclers in Latin America and give them an appreciation for how this historically marginalized population makes valuable contributions to reducing global warming, a role that often goes unrecognized. She introduces us to ways the poor in low-income countries are contributing to the solving of a problem that they have not created in hopes that we will think about the broader issues of economic inequality and social injustice and the ramifications of our consumer societies that demand products whose manufacturing processes—often in developing countries—create environmental harm and waste.

Part III: Going Forward—Emerging Questions and Desirable Evolution in Global Sustainability Risk Management

In Chapter 9, "Managing the Risks of Carbon Sequestration: Liability Concerns and Alternatives," Elizabeth Lokey Aldrich, Cassandra Koerner, Joseph Perkowski, and Travis L. McLing explore one of the possible approaches to mitigating our species' "contributions" to climate change—carbon capture and sequestration (CCS). Focusing on the need to manage the possible liabilities of CCS projects, they discuss how the risks associated with this evolving technology can be understood, assessed, and—perhaps—handled.

The organizations and individuals responsible for developing possible carbon sequestration solutions to our use of fossil fuels must address their responsibility for long-term liability as they simultaneously work to improve performance, reduce deployment schedules, and increase cost-effectiveness. After a brief review of the technical basics of carbon sequestration, the authors examine works in progress in other fields that provide suggestions on how to build on legacy legal arrangements for liability assignment. Experience with such existing regulations as those for subsurface storage of petroleum and natural gas may provide some guidance in grappling with the complex liability situation posed by CCS.

Although collective industry experience, both on similar projects and on incipient sequestration efforts, can provide some guidance on the range of procedures needed to be managed, a comprehensive regulatory framework for effective long-term storage is lacking and represents a significant barrier to CCS projects and initiatives.

The authors identify some of the key questions that call for answers, answers that are yet to be developed. These questions include Who is liable for the CO_2 leakage that might occur long after the establishment of the injection sites? Who is responsible for the ongoing verification effort when CO_2 injection ceases? Since it is possible in some configurations that CO_2 may inadvertently migrate into areas containing resources such as hydrocarbons or drinkable water, how should overall liability (including possible loss of otherwise recoverable resources) be addressed? When the geographic extent of storage sites crosses multiple jurisdictions, how will liability with multiple stakeholders be resolved?

They finish the discussion with a consideration of the larger-level question of whether real-world societal constraints may make implementation of even the most reasonable approach to CO_2 liability management more difficult than it at first might seem to be.

In Chapter 10, "Taking Actions to Deal with Climate Change Risks and Opportunities: Harnessing Superordinate Identities to Promote Knowledge Transfer and Creation," Aimée A. Kane and Alison L. Steele address the crucial question of how businesses, and other organizations, can take the types of actions that are needed to handle the risks of climate change and to move toward a more sustainable world. Building on behavioral science research studies on social identity and knowledge transfer, they identify the importance of how people see themselves in relation to others and to their organizations. They describe how willingness to be attentive to new ideas and information and to consider them carefully can be influenced by one's relationships with the sources of that new information and those new ideas. Building on the social concept of a psychological sense of belonging to an overarching unit that is termed "superordinate social identity," they show how the combination of an organization's clearly espoused commitment to reducing the risks of climate change and contributing to a more sustainable world, combined with organizational members' feeling of identification with the organization, can contribute to people's acceptance of new ways of thinking and behaving—the types of changes needed if companies are to adopt the new technologies and ways of operating necessary to reduce climate risks.

The combination of organizational commitment to reducing climate change and organizational members' superordinate social identification with the organization—seeing and feeling positively about their relationship to the organization, and its mission as a whole—can overcome the "not invented here" syndrome that so often causes individuals and groups to ignore the value of new ideas developed by others elsewhere in the organization or society. The identification with higher purposes, whether in an employing organization, or perhaps with society's higher purposes in general, can help convert the "here" in the "not-invented-here roadblock" into an "invented-here highway."

Kane and Steele emphasize how important it is to reduce the not-invented-here barrier to learning about and adopting valuable ideas for reducing climate change risks because many of those ideas are initially unfamiliar and ambiguous with not easily recognized merits—low in demonstrability of the value of the new approach and with hidden merits. They find that identifying with a superordinate organization contributes to the willingness of individuals to pay attention to and invest time and energy into considering ideas emanating from sources with which they identity and—very importantly—accepting high, but not low, quality ideas from that source, just the kind of behaviors needed to adopt and make happen the many new ways of acting and being required for a more sustainable world.

James A. F. Stoner and Charles Wankel's concluding chapter, Chapter 11, "Final Thoughts," summarizes, very briefly, the editors' final reflections on the overarching themes they hear in the work of the preceding chapters' authors and what that work might imply for all of us.

References

Churchman, C. W. 1967. Wicked problems, Guest Editorial, *Management Science* 14, 4: B141–412.

Estrada, D. 2009. Latin America: The climate clock is ticking. *Tierramérica online.* http://www.tierramerica.info/nota.php?lang=eng&idnews=3277&olt=454 (Accessed January 28, 2010).

Knight, F. H. 1921. *Risk, uncertainty and profit.* New York: Houghton Mifflin Company.

Lovins, L. H., and B. Cohen. 2011. *Climate capitalism: Capitalism in the age of climate change.* New York: Hill and Wang.

Steffen, A. 2006. *Worldchanging: A user's guide for the 21st Century.* New York: Abrams.

Stoner, J. A. F., and C. Wankel. 2009. "The only game big enough for us to play." In *Management Education for Global Sustainability*, ed. C. Wankel and J. A. F. Stoner, 3–17. Charlotte, NC: Information Age Publishing.

Tuck, L. 2009. *Latin America's green path forward.* Tierramérica online. http://www.tierramerica.info/nota.php?lang=eng&idnews=3282&olt=455s (Accessed March 15, 2011).

World Commission on Environment and Development. 1987. *Our common future.* Oxford, UK: Oxford University Press.

CHAPTER 2

Asking the Right Questions: Basic Concepts of Risk Management for Developing Adaptation and Mitigation Strategies

Sujit Sur

We need to provide a third choice between being poor and frugal and being rich and wasteful. And we need to do it very, very quickly.

Alex Steffen, *World changing: A User's Guide for the 21st Century,* 2006

Climate change is a story about the meeting of Nature and Culture, about how humans are central actors in both of these realms, and about how we are continually creating and re-creating both Nature and Culture. Climate change is not simply a "fact" waiting to be discovered, proved or disproved . . . neither is (it) a problem waiting for a solution . . . (it) is the unfolding story of an idea and how this idea is changing the way we think, feel and act. Not only is climate change altering our physical world, but the idea of climate change is altering our social worlds.

Mike Hulme, *Why we disagree about climate change* (xxviii), 2009

Introduction

Climate change has been labeled a "wicked problem par excellence" (Jordan, Huitema, van Asselt, et al., 2010) because of its scientific complexity and lack of policy agreements on how best to manage its impacts. To complicate matters further, those who accept anthropogenic (human-activity-related) causation of global warming and climate change point to human desires for betterment of our quality of life and business/economic pursuits as the root cause for the increased greenhouse gas emissions, while those who propose a natural causation point to a deterioration in the habitability of the planet and the resultant impact on mankind. This state of affairs leads to a peculiar situation where sustainability-oriented climate-change mitigation efforts are seen by many as either resulting in reduction of our present quality of life (becoming frugal), or as a burdensome drag on the global economy (becoming poor), or both—all in an attempt to address an ambiguous issue with an indeterminate outcome and an unknown likelihood of occurrence. Such uncertainties in turn lead to further delays in developing any coherent attempts at addressing the underlying issues, and these delays are seen as exacerbating the issue further as human activities like deforestation and greenhouse gas emissions continue to increase. There seems to be consensus that the "risks" involved are untenable, though perspectives on what needs to be done, how, and by whom are conditional on different factors and sometimes at cross purposes. Thus, to be able to address this wicked problem, we first need to understand the basic premises of the differing perspectives and understandings of risk.

This chapter attempts to clarify basic concepts and differing perspectives on risk and uncertainty. It highlights the fact that our mind-set, i.e., our understanding of risk per se, is contingent on the normative or chosen scientific discipline that we are using to evaluate risk, along with the behavioral, societal, and cultural biases that are implicit with this mind-set. Thus, differences in our mind-sets and decision-making criteria are manifested in the multitudes of normative prescriptions on how to manage the "risks," and what might be the appropriate strategies to mitigate these risks. Understanding these differences in mind-sets is crucial for us to be able to overcome them and to undertake the required transformative changes toward a sustainable world.

Climate change—whether anthropogenic or not—is the largest risk facing humanity and needs to be addressed either by successful mitigation, i.e., reducing the impacts and capitalizing on the opportunities; or by adaptation, i.e., reducing vulnerabilities or losses. And business, because of its centrality to human endeavors and its inherent expertise in risk management, is ideally positioned to undertake this challenge—once the business world understands what really needs to be done and transforms its mind-set.

The chapter is organized as follows: The first section presents the typology of risky conditions as offered by the field of economics. It explains three different conditions related to "risk," their possible outcomes, the decision-making process for each, the consequences of decisions, and the resultant implications for economic gain/loss and resource utilization. The second section brings sociology and behavioral sciences into the discussion of risk.

It elaborates on the perceptual basis and behavioral aspects of response to risky conditions like climate change and the heuristics utilized for decision-making and the underlying mind-set upon which those heuristics are grounded. The third section provides some of the behavioral and psychometric research that bears on and supports the sociological and behavioral perspectives presented in the third section. The fourth section explores the implications of differences in risk assessment and decision-making criteria for climate change and general sustainability-related initiatives, and outlines the possible rationale for the lack of consensus, public apathy, and the increased polarization in the rhetoric. The fifth section develops a framework that integrates the different perspectives and recommends possible strategies for managing risks and opportunities, i.e., mitigating the impacts of climate change or, in the worst case, adapting to the threats of climate change. It underscores the reason why business and corporate leaders are ideally positioned to bring about the requisite transformational change in human interaction with Nature. The sixth section establishes the mind-set required for asking the right questions about climate change and for committing to appropriate mitigation and adaptation strategies. The final section concludes the discussion.

Risk, Ambiguity, and Uncertainty: An Economics Perspective

In economics—the predominant framework for businesses—the differences between risk, ambiguity, and uncertainty are relatively well established and considered to be crucial (Knight, 1921). These concepts differ in their conditions and the opportunities that they present as well as their decision-making criteria, their outcomes, likelihood of occurrence, resultant consequences, and in their economic and resource utilization implications. Alvarez and Barney (2005) elaborate on the distinction between these three concepts as follows:

Risky condition: A market opportunity is considered to be risky when all possible future outcomes of exploiting that opportunity are known at the time a decision is made and when the probability of each of these outcomes occurring is also known at the time of deciding. Thus the outcomes of risky decisions are governed by well-defined probability distributions. Making decisions under risky conditions is analogous to rolling a die known to have six sides and known to be fair and balanced. Although the outcome of rolling the die is not certain, the full range of possible outcomes from rolling this die is known, and the probability of each outcome occurring (1/6) is also known. These characteristics make it possible to calculate a probability distribution that can be used to anticipate the possible outcomes from rolling this die. Most economic and financial models of business decision-making are applicable to risky decisions (Alvarez and Barney, 2005, p. 778).

When faced with such conditions or opportunities, the mitigation or risk-management strategies are to compute the probability distribution of "unfavorable outcome(s) and seek insurance by bundling different opportunities together to make a portfolio, so that overall the organization remains profitable" (Knight, 1921, p. 254). Typically the organization would provide a small amount as a premium to another organization that would pay out if the unfavorable outcome occurs. Such an arrangement is also referred to as financial hedging (Buehler, Freeman, and Hulme, 2008a; Smith and Stulz, 1985) or risk transference. Chapter 3 provides details on the mechanisms of insurance and risk transference.

Another alternative is to hedge operationally; i.e., the corporation diversifies its operations so as to ensure that it remains profitable overall, even if one or a few of the operations are not (Buehler, Freeman, and Hulme, 2008b), or to secure trades for a future date in the futures market, so as to remove uncertainty and volatility from operations. The consequences of such conditions/opportunities and the mitigation strategies employed lead, at best, to increased economic gains for the corporation as it reduces its losses by insuring or hedging against those losses. The corporation reaps the full benefits of the favorable outcome and incurs only the additional costs of the insurance premiums and/or the costs associated with hedging. In the worst case scenario, the corporation might have a series of unfavorable outcomes, not benefitting from the anticipated gains from the favorable outcome and additionally incurring the insurance and hedging-related costs. However, the hedging and/or insurance instruments ensure that the corporation does not suffer the full impacts of the unfavorable outcome, as long as overall "premiums" the company pays are exceeded by the insurance payouts it receives.

The underlying premise of such a risky condition is a "fixed pie" mind-set, a zero-sum logic wherein the economic value is redistributed amongst the players. This condition also presupposes known values and a linear system; i.e., an unending supply of resources (though they might become more expensive), and that the by-products and waste emitted are "externalities," and thus of no consequence (though there might be an increased cost of disposal). However, as long as prices (demand-based) outweigh costs (supply-based), and "risks" are mitigated via transference (insurance) or hedging, the business can generate net "value" and continue operations. Also, insurance and future trades are contingent on at least another player "betting" on the alternative outcome and thus there is a presupposition of acceptance of the transference of risk. There is no additional resource or value created, consumed, or destroyed under this condition, as the (primarily financial) resources merely change hands via the process of arbitrage from one corporation to another (Ghemawat, 2003). Furthermore, systematic risks are not eliminated in this process—they are merely redistributed amongst the corporations. Most corporations with well-established resource bases, especially financial institutions including insurance companies, operate under these premises. They rely heavily on sophisticated statistical analysis and financial modeling to extrapolate trends and compute investment decisions and cost-cutting measures to seek arbitrage opportunities.

Such mitigation strategies are most often undertaken by corporations working in isolation in a competition-based mind-set where each corporation is attempting to uncover and exploit any possible "market" opportunity to maximize its economic gain. This process is fundamentally about maximizing efficiencies. Because the outcomes and their probabilities are known, the corporations are concerned about maximizing their returns with minimal expense. The implications of operating in the risky condition mind-set for resource utilization and value redistribution in the context of climate change and sustainability is discussed in a later section.

Ambiguous condition: A market opportunity is considered ambiguous when the decision makers know, ex ante, the possible outcomes associated with making a decision but not the probability or likelihood of these different outcomes. In the dice analogy, in ambiguous settings, decision makers may know that the die has six sides but they might not know if it is fair or how the die is balanced (Alvarez and Barney, 2005, p. 791). Examples of such a condition might be the "war" over the high definition optical disc format choices between HD DVD and Blu-ray between 2002 and 2008 where Blu-ray emerged as the "winner" ultimately, or the vidoetape format war between VHS and Betamax where VHS emerged as the winner. In case of the optical disc choices, both HD DVD and Blu-ray emerged between 2000 and 2002 and were considered to be equally appropriate initially. The outcome was known because only one of these formats would be established as the worldwide standard. However, the probability or likelihood of acceptance for either of the formats was still unknown and was contingent on adoption of the new standard by other corporations and the alliances that supported either of the formats.

When faced with such conditions or opportunities, the mitigation or risk-management strategies are to seek "toeholds" in as many of the possible outcomes as possible because it is still uncertain as to which outcome will prevail and turn out to be worthwhile. Folta asserts that firms undertake linkages like direct minority holdings and joint ventures, especially when confronted by technological uncertainty, because they provide an *option to defer* internal development or acquisition of a target firm or venture (Folta, 1998, p. 1008). Another mitigation strategy is for the corporations to form alliances with other corporations with complementary capabilities, technology, or resources with the intention of knowledge acquisition or building a knowledge base (Inkpen, 1998). Thus, the mitigation strategies under ambiguous conditions are to undertake call/put options—to form alliances or joint ventures to pool resources and/or R&D to "bet on" the outcome that might eventually succeed. It is important to stress that the decision to choose a particular outcome is taken before the probability of occurrence or "success" is known, and thus the decision is contingent on the judgment, estimation, and ideological preference of the alliance partners.

The consequences of such a strategy in the best case are that the corporations share the research and development costs, thereby reducing each corporation's acquisition costs, augmenting overall capabilities, and reaping the benefits of being the first ones to utilize the new technology or capability (e.g., Blu-ray, VHS). In the worst case, the alliance that "bet" on the outcome that turned out to be unsuccessful (e.g., HD DVD, Betamax) has reduced gains and thereafter needs to provide higher acquisition rent to license or buy the technology from the "winning" alliance. From a resource perspective, under ambiguous conditions, corporations seek value exploration and eventually collaboratively develop and adopt new value, capabilities, or resources while the alternative technology or capabilities are abandoned, leading to some value (and resource) reduction amongst the "losing" alliances. However, there are still some learning and knowledge gains even amongst the unsuccessful ventures, and thus overall there is new value and resource creation from the opportunity for exploratory and collaborative search under ambiguous conditions. It is important to note that some large firms, especially those with immense R&D capabilities, might attempt preliminary exploratory search by themselves; however, it is highly unlikely that they would continue to commit resources equally to two or more likely outcomes for very long.

This process is fundamentally about clarifying effectiveness, as the possible outcomes are known. The corporations are concerned about investigating the full economic potential at optimal costs, prior to focusing on efficiencies. The implications of operating in the ambiguous condition mind-set vis-à-vis resource utilization and value exploration in the context of climate change and sustainability are discussed in a later section.

Uncertain condition: A market opportunity is considered uncertain when the possible outcomes of a decision and the probability of those outcomes are not known when the decision is made. In these situations, decision makers are ignorant of possible future outcomes. Making uncertain investment decisions is also analogous to rolling a die. However, in the uncertain case, the number of sides on the die—whether it be two, three, four, eight, or an infinite number of sides—and whether the die is balanced and fair are not known when the die is rolled (Alvarez and Barney, 2005, p. 779).

When faced with such conditions or opportunities, the mitigation or risk-management strategies are twofold. Entrepreneurs or entrepreneurial firms subjectively perceive an outcome and assess the likelihood of its occurrence on the basis of their ideological preferences, judgments, or estimations and invest substantial resources and effort into developing the initiative. On the other hand, the organizations that do not see the potential consider the efforts and resources to be wasted and either adopt a "wait and see" approach or discount the venture. Alternatively, they actively resist or deny/deride the venture if they perceive their way of working to be threatened by the "innovation" in the unlikely chance of its succeeding.

Those who see the opportunity invest substantial R&D resources and efforts, and those who do not see the potential defer decision or actively resist or deny the initiative. When the entrepreneur's or entrepreneurial corporation's assessments are accurate and they are able to gather the requisite resources and overcome resistance and objections, they are able to create new value or new resources and technology and capture the large-scale economic gains that result (Ardichvili, Cardozo, and Ray, 2003). A classic example is the 30 plus years' commitment by Corning to transform itself from a glass cookware manufacturer to become the world leader in fiberglass optics without knowing the full potential of success, yet ideologically committed to the venture (Miller and Breton-Miller, 2005). In the worst case scenario, the enterprise burns through all its resources either as the goals were unrealistic or as the corporation was unable to muster sufficient resources or overcome resistance (Thornhill and Amit, 2003). The outcome is value depletion, as in destruction of the enterprise, though there might be some learning and knowledge-oriented gains from the process that increase the chances of future success by the individual or surviving corporation.

Thus, depending on the outcome, new capabilities/values are created due to the resultant innovation, though there is resource depletion or destruction for those who do not succeed. This process is fundamentally about developing new standards of effectiveness and the focus is not on efficiencies as they cannot be computed since the possible outcomes and likelihoods are unknown. The corporations are concerned about investigating the full economic potential based purely on their subjective assessment of the associated costs and benefits. The implications of operating in the uncertain condition mind-set vis-à-vis resource utilization and value creation in the context of climate change and sustainability are discussed in a later section.

As can be seen above, these three conditions have critical differences in their possible outcomes and the likelihood of such occurrences. These differences

Table 2.1 Comparative overview of the distinction between risk, ambiguity, and uncertainty

	Number of possible outcomes	*Probability (likelihood) of occurrences*	*Decision-making process for mitigation*	*Consequences*	*Resource implication*
Risk	Known	Known	• Hedge bets	Best: Increased gain; Reduced loss.	Redistribution (Favourable)
			• Insure	Worst: Reduced gain; Increased loss.	Redistribution (Unfavourable)
Ambiguity	Known	Unknown	• Undertake Call/Put Option	Best: Reduced (acquisition) cost; Increased gain	Exploration (Adoption) Value enhancement
			• Form Joint Ventures or alliances to pool resources	Worst: Increased (acquisition) cost; Reduced gain	Exploration (Abandon) Value reduction
Uncertainty	Unknown	Unknown	• Entrepreneurial • Invest in R&D	Best: Maximum gain	Creation (Innovation) Value creation
			• Defer • Resist	Worst: Complete loss	Depletion (Destruction) Value depletion

result in significant differences in the choice of decision-making processes and risk mitigation strategies, the consequences of the decisions, and the implication for economic gain/loss and for resource utilization. Table 2.1 illustrates the main distinctions between these concepts and their implications.

However, while economics and finance scholars generally view the market conditions and resultant opportunities as objective measures with a positivist and near mechanistic flow of outcome and probability-related decision-making, sociology and behavioral scholars generally view all risk-related conditions to be perception-based with socially constructed meaning. Thus, while business decision-making primarily operates on the economics-driven mind-set, individuals, society, and governments operate in behavioral and sociopolitical frameworks. Therefore, except for the shared purpose of generating wealth, more often than not business and society seem to hold divergent views about the associated "risks" of most corporate undertakings. Hence an understanding of risk and uncertainty in a behavioral and sociopolitical framework is necessary for ascertaining the points of commonalities and divergences in these mind-sets.

The next section details the sociological and behavioral perspectives on risk/uncertainty, the decision-making criteria, and the underlying mind-set leading to mitigation strategies, and their consequences and implications for business.

Risk, Uncertainty, and Behavioral Responses: A Sociological Perspective

Research in the sociopolitical field points toward a value-driven and social constructivist view of risk and uncertainty (please refer to Zinn [2008] for an excellent overview of the social theories of risk and uncertainty). In these perspectives, the concepts differ significantly from the economics perspective in their assessment, and the decision-making heuristics applied to mitigate the "feeling of risk" (Slovic, 2010). The basic premise of these perspectives is that risk and uncertainty are social constructions, and thus subjective in the assessment of their outcomes and/or in the likelihood of their occurrence. Some perspectives even assert that risk has become a force for social change (Adam, Beck, and Loon, 2000; Giddens, 1991), and thus the notion of risk is a fundamental part of societal evolution. A synopsis of five perspectives, their decision-making processes and underlying mind-set leading to mitigation or response, and their behavioral and business-related implications follows. The perspectives discussed are: risk society, edgework, systems theory, cultural theory, and governmentality.

Risk Society: This perspective was introduced by Beck (1992) and encompasses a broad societal theory and research prospect that incorporates the paradoxical processes of reflexive modernization and individualization for developing a real yet socially constructed perception of risk (Zinn, 2008). The increasing (real) risks associated with a technologically advanced and industrialized society combine with the lack of identification and containment of these risks to create the central paradox in Beck's theory. The dissemination of information or production of knowledge regarding these risks further replicates them and induces a state of crisis that transforms the order and structures of the world into the risk society. Such a condition is akin to ambiguous conditions where the (real) outcomes may be known, but the likelihood is (subjectively) socially created.

Although these risks and hazards were present in industrialization right from the beginning, it is only after its manifestation in the "limits to growth" debate that these are seen to undermine the institutions of industrial modernity. As long as these risks are secondary to scarcity or needs, the institutions of industrial society will contribute to their proliferation (Beck,

1996). However, somewhere in our relatively recent past, a change occurred in the perception of social order as being based on flows of goods and bads, rather than goods alone. This change in perception has led to a crisis in the way in which modern society organizes and manages its institutions and functions (Loon, 2002, p. 21). Thus society and or its institutions, e.g., governments and businesses, will need to "socially construct" the mitigation of these risks and in the process transform itself into the next (postmodern) social structure and order.

Edgework: This perspective was initially introduced in the context of voluntary risk-taking in extreme high-risk leisure activities. It has expanded to incorporate occupations such as firefighting, police work, criminology, search and rescue, test piloting, combat soldiering, and even to treading stocks and bonds in the finance industry (Lyng, 2008). The wide range of empirical phenomena offered in the different accounts relating to these risky activities represents a unique framework for the study of risk and uncertainty within a predominantly positive perspective and provides important insights into the problems of risk and uncertainty in the contemporary social order. However, the perspective has yielded very little in the way of testable propositions relating to decision-making in risky situations (Lyng, 2008, p. 135). The individuals in these activities are very much like entrepreneurs operating under uncertainty without a clear idea of the outcomes and likelihood beyond their subjective judgment and experience-based estimations, and the decision-making processes of the two groups are similar. Edge workers, like entrepreneurs, typically develop innovations and/or lead changes, though mediated by society, and sometimes create value for themselves and/or for society.

Systems Theory: This perspective is an analytical approach that bases itself upon the notion of modern societies' structures, which are shaped according to the action-function of Talcott Parsons (1980) or are visible in the form of the communication-function (Luhmann, 1982; Luhmann, 1995) of different specialized subsystems. In terms of conceptualization of risk, systems theory offers a unique approach by placing risk at the center of its notion of modern society (Japp and Kusche, 2008, p. 83). By definition, systems theory implies that modern societies depend on decision-making in relation to risk as a fundamental fact, and while it does not allow for a hierarchical top-down and stratified decision-making process, it relies on a distinctly modern, functionally differentiated, lateral, and self-referent decision-making process. Systems theory allows for conflict. Rather than seeking consensus through communication, it looks toward pragmatic assumptions; i.e., abstaining from any attempt at real or authentic understanding of difference as a basis for discourse. The typical result is that mutual understanding is simply assumed, and in this way it has actual consequences. Distinctions enabling the consideration of other perspectives are needed, even if only in a tentative and provisional way (Japp and Kusche, 2008, p. 92). Thus risks are considered objective and real, though the assessment of outcomes, likelihood of occurrence, and the choice of mitigation strategies are a collective (subjective) process based on action functions and communications among the different actors. This process inherently is protracted, discursive, and emergent in nature. All the actors (institutions, society, and even nation-states) need to participate and contribute to the actions and communications. Although the resolution is generally effective and comprehensive if and when completed, the timeliness and even the probability of completion cannot be known ex ante.

Cultural Theory: Even before the notion of climate change intersected with the notion of large-scale risk, Douglas and Wildavsky (1982) developed a theory focusing on perceptions of pollution and other ecological disasters as being a boundary defined in terms of social, psychological, and cultural conditioning. The overarching framework of Douglas and Wildavsky's Cultural Theory offers a classification of four types of individuals distinguished by their group-oriented or individual-oriented perspectives, as well as their behavior control preferences and their "ways of life" (Hulme, 2009, p. 186). These typologies and their associated perceptions and mind-set scan be mapped on the four classifications of Buzz Holling and Michael Thompson's "myths of nature" (Hulme, 2009, pp. 188–190). These four classifications and their perception of nature are as follows:

- *Hierarchists* have a high degree of social regulation and social contact and see nature as perverse but tolerant if treated with care. They assume that the climate system is uncontrollable to a certain degree, but is quite resilient if suitably managed. They see the risks to be real and foresee the need for more knowledge of climate systems and predictive capabilities to be able to develop appropriate mitigation strategies, which would ensure that the outcomes of climate change are managed to be sustainable (Hulme, 2009, p. 190).
- *Egalitarians* share a high degree of social contact but a low degree of social regulation and see nature as ephemeral and existing in a precarious and delicate state of balance. They accept the premise that the slightest perturbation by humanity can trigger a collapse in the system. They tend to foresee the risks of climate change to be frightening and easily spiraling out of control. They believe that the mitigation strategy needs to involve

altruism and common effort, as in voluntary reduction in emissions in an attempt to maintain the status quo (Hulme, 2009, p. 190).

- *Individualists* share a low degree of social contact and a low degree of social regulation and see nature as benign and foresee the climate system (including "normal" changes) as favorably inclined toward humanity, at least within broad definable limits. Any risks due to climate change are viewed as manageable, as the earth's climate will reestablish itself at a tolerable and nondangerous level. Such individuals view future outcomes as personal, individual responsibilities and prefer to adapt and change in the future, if required, rather than rethink present technological processes (Hulme, 2009, pp. 189–190).
- *Fatalists* share a low degree of social contact but have a high degree of social regulation and see nature as capricious and view climate change as fundamentally unpredictable, influenced by multiple factors—of which human activity is just but one. Such individuals view the risks with apathy and a sense of fatality as Nature has always been a risk for humanity and will continue to be one (Hulme, 2009, p. 190).

Governmentality: This alternative analytical approach, proposed by Foucault, is an attempt at understanding politics and power. In particular, this approach seeks to reflect the multiplicity of power throughout various institutions that adhere to the same governing mentality while maintaining a balance with self-governing capacities of the different individual and collective entities (institutions). Governmentality is essentially a practical, rational, and technological mentality in which the conceptualization of risks is viewed as a calculative technology of governing (O'Malley, 2008, p. 57). Within this perspective risk factors or pools can be viewed in statistical and probabilistic terms, which thereafter produce numerically based predictions and enable governing and decision-making processes. However governmentality is not a consistent set of concepts and theorems developed for explanatory purposes, rather it is a heuristic analytic and does not pretend to be all encompassing (O'Malley, 2008, p. 68). Such a mind-set is consistent with the risky condition explained earlier, where the possible outcomes and likelihoods are computed to develop mitigation strategies. However, as the aspirations of individuals and collective entities are values-based, they defy easy computation. Furthermore, when differences emerge between analytical and experimental data, the differences create a dilemma for policy makers. On the one hand, they are urged to follow deliberative approaches (e.g., cost-benefit analysis) that act as a check against unwarranted fears (Sunstein, 2005), but, on the other hand, they are also advised to respect the public's sensitivity to important value-laden considerations that are often ignored in expert deliberations. Thus, in spite of its focus on calculative technology, governmentality-based risk assessment and management ultimately remains a socially constructed perspective, with concerns similar to those of the systems theory perspective.

As can be seen above, each perspective on risk operates in its own discipline-based paradigm and within its institutional or societal context, in independent though interrelated mind-sets. What is of importance here is to understand how the different actors respond to their perception of risk, whether subjective or objective, and what decision-making heuristics are utilized to develop their responses. Research from the behavioral sciences, especially psychometric studies, holds important findings that can not only enable understanding of the present mind-sets, but may also provide the key to how best to transform these mind-sets to be able to address issues of sustainability. A quick overview of some of the important findings is in the section that follows.

Psychometric Perspectives on Behavioral Responses to Risk and Uncertainty

People apprehend reality in two fundamentally different ways—emotive/subjective andanalytic/objective (Epstein, 1994). Reliance on emotions and intuition seems more efficient to most people. The analytical perspective is resorted to when effectiveness is in doubt. Slovic (2010) builds on this premise to develop the affect heuristic approach to explain the *feeling of risk*: the notion that our perception of risk, acceptable risk, and perceived benefits relate to intense feelings of dread associated with particular hazards (Slovic, 2010, p. xxi). These feelings serve as an important cue for risk/benefit judgments and decisions. If we like an activity, we tend to judge its benefits as high and its risks as low; if we dislike it, we judge it the opposite way, low benefits and high risk. In the case of risks like climate change, which require technological and quantitative data and lack visceral, emotional experience-based meaning, these types of risks may be too amorphous to create any opportunity for transformation unless they are framed in a more coherent fashion (Slovic and Slovic, 2010, p. 81).

The challenge or risk of climate change in a democratic/cultural context represents another level of complexity. Sunstein's (2005) concepts of "risk panics"—where members of the public form widely overstated estimates of some societal dangers and curiously understated estimation of others. Sunstein's (2005) model of expert cost-benefit analysis, when properly framed in terms of affect heuristics, might help separate out considered public values from irrational public fears or the "irrational weighed model" (Kahan, Slovic,

Braman, et al., 2010, pp. 184–185). However, questions of numeracy skills and public understanding of perceived probability and utility can further impact the affect heuristic and therefore alter any decision-making process in peculiar ways (Slovic, 2010).

In addition to the content of the situation, the framing and context of communications are also critical for appropriate decision-making. And, once established—whether appropriate or not—these effects continue to impact our subsequent decision-making (Ariely, 2010) via *imprinting.* This impacting occurs because our learning and experiences lead to neurologically positive and negative feelings, "markings," of somatic or bodily states (Damasio, 1994). These feelings trigger our reaction to a future outcome. A negative marker triggers resistance, while a positive marker triggers incentive. When risk events undergo substantial amplification, they may result in unexpected alarms or "social shocks." At the other extreme, risk attenuation may occur. Despite the serious consequences to the bearers and society in general, risk events may pass virtually unnoticed and untended, often continuing to grow until reaching disastrous proportions (Slovic, 2010, p. 325). The scientific community's highly technical and impersonal framing of explanatory communication does not generally allay such fears or mitigate the impacts of risk attenuation.

The importance of framing for response is well established in terms of prospect theory (Kahneman, Knetsch, and Thaler, 1990; Slovic, 2010; Tversky and Kahneman, 1974). A focus on gains or losses generates different *preferences,* and some individuals tend to prefer certain to uncertain alternatives (Kahneman and Tversky, 1979). This preference is termed the *certainty effect.* Prospect theory also contends that individuals are more sensitive to minor losses than to minor gains, and even the perception of loss or gain depends on the *reference point* (Tversky and Kahneman, 2004)—a subjective estimation, generally based on comparison either with others or with previous experience.

Thus, the very definition or assessment of risk, the decision-making criteria, the understanding of its outcomes, and the likelihood of occurrence are contingent on the perspective of the actors and their mind-set. Most people presently view sustainability initiatives as unwieldy costs, or as resulting in a reduction in quality of life (Dyer, 2008). In the sociological perspective, individuals' responses to climate-related risk and uncertainty fall into three distinct categories: denial, apathy, or transformation. This classification maps rather well on the evidence on public response toward the risks associated with climate change. Six categories of public perceptions and preferred responses to climate change were found in a US study (Leiserowitz, Maibach, Roser-Renouf, et al., 2010) and four categories in a UK study (Langford, 2002).

These findings are very much in line with cultural theory. Our ability to address the real or perceived risks of climate change will need a rethink of our accepted norms about the role of society and nature. However, promoting sustainable development within organizations and society as a whole presents more than just economic, social, and environmental challenges. Public enlightenment is mediated by the cultural typologies; i.e., hierarchists, egalitarians, individualists, and fatalists. And individuals are likely to select information in a biased fashion. Nations and socioeconomic segments of society also interject their preferences based on either their need for *certainty* or their apprehensions about perceived loss in quality of life in *reference* to those better off. Faulty assumptions about the nature of the opposition in ideologically based disputes based on biases and mind-sets are created and perpetuated by institutions. These interrelated contextual mechanisms—individual values and social institutions—need to be taken into account to understand failures in ideologically based negotiations (Wade-Benzoni et al., 2002). These structural differences in perception are the underlying rationale for the lack of consensus, the polarization in the rhetoric, and the general public apathy on such a vital issue (Hoffman, 2011; Hulme, 2009).

Apart from these societal, national, and international discourses, climate change risks also present challenges in terms of consciousness development and the promotion of new action logics within corporations (Boiral, Cayer, and Baron, 2009). Since the primary focus of this chapter is on business corporations and businesses have the most control on eliminating or reducing human activity related to greenhouse gas emissions, the subsequent discussion is limited to the implications for corporations.

The next section attempts to integrate all of the above concepts and draw out the implications, especially in terms of business's response to climate change and for business's role in general sustainability related issues.

Implication for Business Decision-Making on Climate Change and Sustainability

Whether climate change is caused by business, unrelated human activities, or natural causes, the concerns need to be urgently addressed. The globe is warming and greenhouse gases do contribute to that effect. The scenarios of "climate wars" based on Dyer's (2008) interpolation of events, highlight the grave consequences of inaction (Craven, 2009). The political process, though in some agreement about the need to reduce greenhouse gases, is gridlocked over distributive negotiations (Thompson, Wang, and Gunia, 2010) and is constantly resetting emission-reduction goals on a sliding scale basis. Meanwhile, the latest report (International Energy Agency, 2010) indicates that energy-related CO_2 emissions in 2010 were the highest ever, reaching 30.6 Gigatonnes (Gt). This level of emissions means that achieving the 2020 target of emissions not greater than 32 Gt per year as set in Cancun in 2011 requires that emissions rise less over the following ten years in total than they did between 2009 and 2010. Such a task is nearly an impossible one, especially without coherent and convincing ways to define *how* to tackle reductions without impacting the economy negatively.

Paradoxically, the actor best suited to take effective and timely action is business, which has been the largest perpetuator of greenhouse gas emissions. Since business is the central actor in climate change and because of its inherent action orientation, it does not have the option of responding with apathy. It must respond either with denial/resistance or with transformation. The key task, therefore, is to achieve the voluntary transformation of business' dominant discourse.

Attempts at planned state/governmental "intervention" usually lead to decoupling business discourse from its activities, and hence businesses might continue their practices according to the dominant discourse but use the rhetoric of the challenging discourse (Ahlstrom, 2010), or resist by lobbying to safeguard their "normal" way of operations by funding the denial groups. In other words, the first reaction of business toward government-mandated sustainability standards might be green washing; i.e., ceremonial/superficial public relations exercises, deferment, or resistance tactics with no substantial alterations of core operations. Imposition of carbon taxation might lead to the adverse impact of encouraging businesses to compute the costs/benefits of operations and resort to cost/risk transference mechanisms—without any ethical concerns as they would feel as though they already "paid" the price for polluting (Ariely, 2010). Hence unless businesses change their underlying "risky condition" mind-set voluntarily, external pressures by stakeholders—even if unified—might not be timely enough to be effective.

Businesses' probable responses to the threats of climate change might involve ignoring the situation or being opportunistic, collaborating, being entrepreneurial, or proactively transforming their mind-set.

1. Ignore the situation or respond opportunistically (businesses adopt a risk mind-set). The largest danger of remaining in the "risky condition"-based framework is that there will be no mitigation of the climate change factors, even if redistribution of wealth favorable to business might continue. This outcome can occur if another entity is willing to take on the risk transference or customers continue to accept the costs being passed on to them. However, such wealth redistribution

is also not an equitable process as the populations in the lower socioeconomic strata suffer disproportionately (Ash and Boyce, 2011) even in the developed economies (McLeod et al., 2000). Such redistribution raises issues of environmental justice. Furthermore, even if some countries or groups are able to absorb the resultant consequences of extreme climate conditions to some degree, others—the nonprivileged and socially vulnerable—will experience the collapse of societal order and escalation of violence (Beck, 2009). Thus if businesses continue to operate in the risky condition mind-set; i.e., make no attempts to reduce greenhouse gas emissions, the "best case" scenario will lead to increasing polarization in society based on unbalanced resource distribution, and survival of the few "haves" and annihilation of the "have-nots." The survival of the few "haves" may not even occur if there is a breakdown of law and order. In the worst case scenario, there will be unprecedented chaos, large-scale migrations to urban and developed nations, and breakdown of law and order and infrastructure, resulting in the survival of the very lucky or perhaps unlucky few (Dyer, 2008).

2. Collaborate (businesses adopt the ambiguity mind-set). This response might be a preferable one, provided that the underlying premise is not oligopolistic and opportunistic behavior is avoided. As Beck (2010, p. 259) states, climate change releases a "cosmopolitan imperative"—cooperate or fail. By joining force to form alliances and joint ventures, businesses can pool their expertise and R&D resources to experiment and explore avenues for sustainable operations on larger scales and collectively reap the benefits of the new technology or means of production. Large corporations can also provide resources to new initiatives thereby taking an "option" on acquiring the resultant technology or benefitting from the knowledge transfer (Busch and Hoffmann, 2009). In the best case scenario, such a collaborative approach might result in coordinated and efficacious results that lead to reduced or reversed climate change and sustainability-driven development. In the worst case, if the alliance partners lack trust or operate opportunistically seeking undue advantage, the outcome would be a stalemate and deferment of any efforts till they might become ineffective.
3. Entrepreneurial (businesses adopt the uncertainty mind-set): This response is also a preferred one, as new businesses (entrepreneurship) or new ventures within corporations (corporate entrepreneurship) start with an ideological mind-set that sees potential in uncertain and dynamic conditions and does not need to conform to the existing (pollution-generating) ways of operating. Aragón-Correa and Sharma

(2003) found that in industries with a high level of technological dynamism, companies tend to take greater risks because there are better opportunities to benefit from innovation. Other researchers also find that industry and firm-specific factors, namely, technological dynamism, complementarity between new technologies and existing assets, and ownership of specialized assets for commercialization influence how companies strike a balance between the different trade-offs and deal with the uncertainty created by the current "climate policy deadlock" (Pinkse and Kolk, 2010). In the best case scenario, such enterprises might develop coherent and effective sustainable initiatives, which thereafter can be rapidly scaled via other corporations to reduce/reverse climate change. In the worst case situation, if no coherent modality is discovered or innovated, the overall results might be fragmented, ad hoc, and ultimately ineffective responses that might be too little and too late.

4. Proactive multimode (all businesses transform their mind-set). This outcome is the ideal one, with all businesses realizing the long-term impact of continuing to operate under the premise of mitigating "risky conditions" and voluntarily and whole heartedly adopting a proactive environmental strategy as a central tenet of their operations. Doing so would involve developing a different set of capabilities and forsaking traditional modes of operations. Such actions would be a net benefit to the corporation, as developing absorptive capacity (organizational capabilities) leads to both competitive advantage as well as proactive environmental performance when upper management is driving the process (Delmas, Hoffmann, and Kuss, 2011). All that is required for this transformation is that business gives up its need for *certainty* and accepts or learns to operate under ambiguous and uncertain conditions (Pinkse and Kolk, 2010). The outcomes will be similar to those presented in the best cases in 2 and 3 above; however, the scale of operations will be much larger and thus the results will be quicker and more widespread.

Table 2.2, adapted from Zinn's (2008) overview of risk epistemology, integrates the different disciplines and approaches to reflect the commonalities and points of departure for all perspectives and summarizes the main points of this section.

Although presented as four different possible business responses, none of the perspectives is discretely unique.As even the most rigorous statistical analysis still reflects past occurrences, any interpolation into the future is an

Table 2.2 Integrative overview of mitigation strategies contingent on mindset and possible consequences

Threat/Opportunity is considered to be . . .	*Perspective*	*Discipline*	*Approach*	*Mitigation strategy*	*Possible consequences for climate change and sustainability*
Real, known and objective	Risky condition	Classic Economic theory (Finance)	Objective calculation of outcomes & likelihood for opportunistic individual profit seeking	• Insurance • Hedging • Future trading • Adaptation	Best: Resource redistribution (short term) to few, annihilation of most Worst: Unprecedented chaos and massive failures, lucky few survive
Real, knowable yet unclear; Socially constructed	Ambiguous condition	Risk society; Neo-economic theories	Objective calculation of outcomes, collaborative (subjective) assessment of likelihood, collective gain seeking	• Alliances/JV to pool R&D • Experimentation • Transformation • Call/put options	Best: Coordinated, efficacious reduced or reversed climate change, sustainability driven development Worst: Stalemate, deferment to 'the point of no return'
Unknown and unclear; Socially mediated	Uncertain condition	Edgework; Entrepreneurial opportunity seeking	Individual (subjective) assessment of outcomes & likelihood, individual gain seeking	• Experimentation • Invest in R&D • Transformation • Resist or Deny	Best: Coherent, effective sustainable initiative development, rapid scaling to reduce/reverse climate change Worst: Fragmented, ad hoc and small scale responses
Subjectively biased	Any	Psychometric paradigm; Behavioral theory	Individual or collective (subjective) assessment of outcomes & likelihood of perceived risks	• Consensus building • Reframing communications • Resist or Deny	Best: Coherent, effective sustainable initiative development, rapid scaling to reduce/reverse of climate change Worst: Fragmented, ad hoc and ultimately ineffective responses

(*continued*)

Table 2.2 Continued

Threat/Opportunity is considered to be . . .	*Perspective*	*Discipline*	*Approach*	*Mitigation strategy*	*Possible consequences for climate change and sustainability*
Socially constructed	Any	Systems theory	Collective (subjective) assessment of outcomes & likelihood of objective risks	• Pragmatic consideration • Provisional agreement • Defer • Resist or Stymie	Best: Coordinated, efficacious reduced or reversed climate change, sustainability driven development Worst: Stalemate, deferment to 'the point of no return'
Socially transformed	Any	Cultural theory	Real threats are transformed into risks for sociocultural boundaries	• Transformation • Technocratic adaptation • Resist or Deny • Apathy	Best: Coordinated, efficacious reduced or reversed climate change, sustainability oriented societal transformation Worst: Fragmented, ad hoc and socio-culturally bound responses
Socially constructed	Any	Governmentality	Events are risks insofar as they are part of a calculative technology	• Negotiating • Posturing • Defer • Resist or Stymie	Best: Coordinated, efficacious reduced or reversed climate change, sustainability oriented planned development Worst: Stalemate, deferment to 'the point of no return'

educated estimation at best and thus subjective. Also when dealing with conditions like climate change and global warming, knowledge of all possible outcomes is next to impossible. Thus the objective demarcation among uncertainty, ambiguity, and risk is largely subjective and perception-based. Although the fact of climate change is undeniable, all of the causes are not known with certainty and the outcomes and likelihood of occurrences are unknowable. Thus the default mind-set needs to be the uncertainty mind-set, and if and whenever that mind-set becomes widespread all that is needed is a critical mass to realize the folly of the other modes of operating. As just one organization succeeds in the transformation, others will follow rapidly. The section that follows highlights some of the drivers for such a process.

An Integrative Framework for Sustainability and Mitigating Climate Change

Business is unquestionably the most powerful institution of modern times. It is best suited to rapid transformation, either via attrition or reinventing itself. It is an institution that is efficiency-driven. Once the direction and objectives (effectiveness) have been chosen, business has had an established track record of achieving its objectives. Undoubtedly many businesses will fail in the process of transformation, but it takes just one to succeed, and others will rapidly build on the success. Change in "business as usual" is an undeniable imperative as human impact on earth is of such magnitude that scientists have already named the period post-2000 AD as the Anthropocene—the new era in which humans rival nature in their impact on the environment (*Economist*, 2011).

The required change has to be in the mind-set with which business approaches climate change and other "externalities" and transforms its operations to a sustainability-driven imperative—without having to forsake its tools and operating principles. As is elaborated further in Chapter 4, sustainability in fact can become the next source of competitive advantage for business, with a relatively minor transformation in its mind-set. In fact, with the required transformation; i.e., reorienting of its mind-set about sustainability-related conditions; business might be the first and best suited to mitigate climate change and establish sustainability as a central tenet of future human enterprise. An illustration of this possibility is the excerpt from William McDonough's talk at the 2005 TED (Technology, Entertainment, Design) conference that follows explaining why his "Cradle to Cradle" firm utilizes the tools of business (commerce) for their sustainability by design initiatives, why there needs to be a change in mental framing, and how commerce will ultimately save the world:

> Commerce . . . is relatively quick, essentially creative, highly effective and efficient, and fundamentally honest, because we can't exchange value for very long if we don't trust each other. So we use the tools of commerce primarily for our work, but the question we bring to it is, how do we love all the children of all species for all time? And so we start our designs with that question. Because what we realize today is that modern culture appears to have adopted a strategy of tragedy. If we come here and say, "Well, I didn't intend to cause global warming on the way here," and we say, "That's not part of my plan," then we realize it is part of our de facto plan . . . it's the thing that's happening because we have no other plan . . . Then, the question becomes not growth or no growth, but what do you want to grow?
>
> So instead of just growing destruction, we want to grow the things that we might enjoy . . . where we take materials and put them into closed cycles . . . Our first product was a textile where we analyzed 9,000 chemicals in the textile industry. Using those intellectual filters, we eliminated 8,762. We were left with 38 chemicals . . . (to develop) the first fabrics (for) infinitely reusable carpet . . . nylon going back to caprolactam back to carpet . . . the (waste) water coming out was clean enough to drink . . . And this is our project for Ford Motor Company. We saved Ford 35 million dollars doing it this way, day one, which is the equivalent of the Ford Taurus at a four percent margin of an order for 900 million dollars' worth of cars. It's the world's largest green roof, 10 and half acres. This is the roof, saving money, and this is the first species to arrive here. These are killdeer (birds) . . . We're developing now protocols for cities . . . We're doing 12 cities for China right now, based on cradle to cradle as templates. Our assignment is to develop protocols for the housing for 400 million people in 12 years. We did a mass energy balance—if they use brick, they will lose all their soil and burn all their coal. They'll have cities with no energy and no food. We signed a Memorandum of Understanding . . . for China to adopt cradle to cradle. Because if they toxify themselves, being the lowest-cost producer, send it to the lowest-cost distribution . . . Walmart . . . and then we send them all our money, what we'll discover is that we have effectively is mutually shared destruction. (McDonough, 2005).

The question that business needs to rephrase is not how can it operate more efficiently, but how can it achieve growth without depleting natural resources (effectiveness) and without producing wasteful (inefficient) externalities? However, once corporations become aware and accept the need, or recognize the entrepreneurial opportunity of developing new action logics and overcoming challenges, they will realize that doing so is not something alien to how businesses deal with emergent issues. In fact, there is a well understood social issues life-cycle perspective (Ackerman, 1975) and a corporate response framework (Sethi, 1979) that illustrate the four stages of business response to social issues: (1) the preproblem stage; (2) the problem-identification stage; (3) the remedy and relief stage; and (4) the prevention stage.

Until the late 1990s, most companies focused more on political, non-market strategies, usually to oppose upcoming regulatory regimes as the benefits did not seem to justify the costs (Bansal, 2002). When corporations were forced to conform to an imposed regulatory standard, they tended merely to adopt ceremonial behavior intended to present an appearance of conformity, or at best they made administrative and technical improvements (Boiral, 2007). Such responses are typical of the preproblem stage where either the existence of the problem is denied or ignored as being too small or transitory to be considered.

Corporations of late have started to look at their operations with a sustainability perspective and to identify and compute their carbon footprints (Dunn, 2002; Islegen and Reichelstein, 2009). Such responses are typical of the problem-identification stage, where corporations take cognizance of the issue and attempt to understand its scope.

Currently, however, a range of market responses is also emerging to address global warming and reduce emissions through product and process improvements, and through emissions trading and other market strategies. Political activities also continue to play a role (Kolk and Pinkse, 2005). These responses are representative of the remedy and relief stage where corporations start attempting to address the issues and seek remedial measures. In accordance with Sethi's framework, businesses are presently poised to enter the fourth and final, preventative, stage, where the corporation transforms itself so as to prevent or reduce the occurrence of the issue.

According to Beck, climate politics is precisely *not* about *climate* but about transforming the basic concepts and institutions of industrial, nation-state modernity (Beck, 2010, p. 256). Until we achieve this transformation, overcome the industrial-waste-induced crisis, and transit from the present period of industrial modernity to the next, we will remain in the present interregnum of the (world) risk society (Beck, 1996). However, favorable resolution of the present crisis is not a given. In fact, Tainter (1988) argues that the historical collapse of sophisticated (complex) societies was precisely because the elites (decision makers) of the society continued to build on the orders and structures that initially helped create surpluses for that society. Even when those structures became unwieldy and cost-ineffective, the society continued to add another layer of complexity to the structure, thereby further exacerbating the conditions instead of simplifying the structures to support the new imperatives of development. Organizations tend to buffer their technological core (Thompson, 1967) and attempt to maintain the status quo due to *imprinting, certainty*, and *reference point* biases (Tversky and Kahneman, 1981). However, business is also accustomed to rapid transformative change when it sees a better opportunity and speedily adopts a new direction.

The section that follows focuses on what might be the right questions that business needs to ask of itself to enable the required transformation to the sustainability-driven mind-set, and how our understanding of decision-making heuristics can aid in the transformation.

Asking the Right Questions: The Required Mind-Set for Mitigation

Businesses are socially mandated to improve the quality of life for all and are encouraged to make a profit as a means for continuing to do so. However, this profit-seeking *means* is often seen as a hyper-efficient, single-minded focus on a profit maximizing *end*, a view supported by the many instances of "profit at any cost" mentality reflected by the Enrons and Worldcoms. Enron, in fact, is a perfect example of the dangers of operating in a risky condition mind-set. It started as a relatively small-scaled energy provider and then the top management computed greater gains (lesser risks) by changing into an energy trader that could transfer all its (operational) risks on to other firms by complex derivatives or by trading in futures markets. Then it began massaging its accounts to capture the financial benefits from exercising stock options with its skyrocketing stock prices. It is easy to attribute these behaviors to individual greed; however, they can also be viewed as a recurrent systemic issue that can be understood in terms of individuals and businesses operating under a risky condition mind-set and in accordance with their escalating *reference point* focus on avoiding the prospect of declining from their previous reference point (Mishina, Dykes, Block, et al., 2010).

Businesses are efficiency-driven, and managers are *imprinted* to derive maximum efficiencies and gains from their operations. Instead of attempting to reprogram and battle with the instinctive resistance to their "way of being," the question might be rephrased to "how can waste inefficiencies be effectively eliminated from operations?" With this phrasing, businesses might be more amenable to applying themselves to realizing the "gain-seeking" opportunities in better resource utilization, instead of resisting due to the "loss avoidance" mentality of incurring additional costs for sustainability-oriented initiatives.

An example might be Ray Andersen, Founder and CEO of Interface, who had an epiphany in 1994 when he read *The Ecology of Commerce* by Paul Hawken (1993) and *Ishmael* by Daniel Quinn (1992) and realized that the industrial system is destroying the planet and only industry leaders are powerful enough to stop it. Ray challenged his company to adopt a bold vision that required new thinking and a new model for business (Anderson, 1998). Since 1995 the company has reduced its waste by one-third, recycles all its products, and plans to have a zero ecological footprint by 2020. Such a transformation comes only with a reframing of mind-sets to include Nature in normal operations, and the onus for the consequences needs to be seen at a personal level, such as in impacting the future well-being of *our* children. Beck also emphatically rejects the notion that society and nature are separate and mutually exclusive, and argues for a paradigmatic shift in perspective from an "either–or" to a "both–and" mind-set and a rephrasing of the question to "How to create a greening of modernity?" (Beck, 2010, p. 258).

Thus there is a need for transforming business's mind-set from the risky conditions mind-set to the entrepreneurial and exploratory mind-set of uncertain and ambiguous conditions. This shift is appropriate because the outcome as well as the likelihood of occurrence for "green modernity" is uncertain and unknown. The situation requires an innovation-based approach. Some examples of such approaches are as follows:

Transformation: Innovation-based approach for sustainability (mitigating the impact). The sustainability and climate change conversation needs to move from a loss frame to a gain frame (Tversky and Kahneman, 1981). Utilizing affect heuristics with mental images of winning creates positive affect that motivates choice (Slovic, 2010, p. 29) and reduces the fear of uncertainty. Business needs to reformulate its structuring from a linear system to closed loop "cradle to cradle" thinking. The emergent concept of biomimicry (Benyus, 2002) provides a framework for time-tested symbiotic design of industrial systems. The focus needs to be on the "uncertain conditions" framework based on upside gains for entrepreneurial initiatives. An additional benefit for business is that employees feel that they are doing something important and making acontribution (Ariely, 2009), thereby increasing productivity and spurring further innovation. Such proactive approaches (Boiral, 2006) to sustainability are akin to undertaking preventive measures, such as lifestyle changes to safeguard health. Other examples of current business models are also readily available (Kolk and Pinkse, 2004, 2005; Lempert, Scheffran, and Sprinz, 2009). Ideally, businesses need to undertake multiple approaches (NRC, 2010c) and create collaborative ventures based on the "ambiguous conditions" framework.

For those businesses with limited resources or that prefer waiting until they can figure out the emergent trend, there still is an opportunity in working toward eco-efficiencies by reducing their emissions or using resources more efficiently as follows:

Dampening climate change effects—while lowering costs (reducing the impact). Businesses can achieve cost savings from energy efficiencies and reduced effluents (NRC, 2010d). Such an undertaking is akin to "regulating" a dietary regime for maintaining one's health. Businesses can switch over to different alternative and renewable energy sources or join a smart-grid to minimize power consumption from power plants because their operations account for the largest amount of greenhouse gas emissions. However, businesses need to be cognizant of the fact that mere reduction is not sufficient as overall consumption is constantly trending upwards, and the time lags for the impact of the reductions might not suffice for bringing climate change to an end (NRC, 2010b). These initiatives are well underway, as businesses have been voluntarily undertaking greenhouse gas emissions reductions as strategic measures, to be prepared for the long term incase reductions become mandatory, while simultaneously reaping near-term economic and strategic benefits (Hoffman, 2005).

Finally, in case the political stalemate continues and businesses do not undertake mitigation or reduction efforts in a timely manner or some other unforeseen events trigger extreme climate change, business still might have an important role in spearheading the adaptation efforts as follows:

If all else fails: possible adaptation strategies if efforts are "too little, too late." Such efforts are akin to intervention; i.e., surgery or medication to keep a patient from dying. These strategies carry an additional risk of unintended consequences, as any technological solution utilized to adapt to a condition might result in other unforeseen complications or interactions. Businesses, like military forces, are well suited to develop contingency planning and strategies, including establishment of clear objectives and opportunities to incorporate adaptation plans into existing management goals and procedures. Businesses also have the ability to identify cobenefits associated with adaptation measures, and the presence of strong leadership (NRC, 2010a) to make a serious attempt to ensure survival (Mendelsohn, 2000; Nitin, Foster, and Medalye, 2009). Some examples might be geoengineering solutions like firing sulphides or deploying reflective surfaces in the stratosphere to reduce the incoming solar heat and thereby reducing surface temperatures. Another example would be carbon sequestration (Islegen and

Reichelstein, 2009) as discussed in Chapter 9. However, these approaches, as of now, might best be considered as options of last resort as the efficacy of such measures is unknown, and deploying them might occur too late to be effective or create new, unforeseen problems.

Conclusion

In conclusion, despite the differences in our perceptions of risks related to climate change and the lack of consensus on what might be appropriate mitigation strategies, business is best positioned to deal with these issues provided it transforms its mind-set. The risks or uncertainty associated with climate change or sustainability-oriented initiatives are not necessarily debilitating for business, nor should such efforts be seen as a compromise, or a Luddite reduction in quality of life, or even as incurring additional cost. Sustainability-oriented operating and managing need to be rightfully seen as the new directive for effectiveness and approached with the requisite entrepreneurial zeal.

The associated investments need to be viewed as the seed capital that will pay back multiple times with the latest series of innovations. For those corporations that are already implementing holistic, sustainability-driven initiatives, the "green dividend" of not wasting anything is quickly understood and adopted system-wide (Farzad, 2011, p. 72). It is leading to increased efficiencies (McDonough and Braungart, 2002) and a revitalization of entire operations (Anderson and White, 2011). These exemplar corporations will be the benchmark for the corporations that are still sitting on the fence. In fact, sustainability-oriented value creation and exploration just might be the spur required for business to reframe its mind-set, transform its operations to eliminate or recirculate waste, and innovate itself to lead mankind in the already unfolding Anthropocene era where business can meet the needs of the present while enhancing the ability of future generations to meet their own needs. As the *Economist* (2011) eloquently summarizes:

> For humans to be intimately involved in many interconnected processes at a planetary scale carries huge risks. But it is possible to add to the planet's resilience, often through simple and piecemeal actions, if they are well thought through. And one of the messages of the Anthropocene is that piecemeal actions can quickly add up to planetary change (*Economist*, 2011, p. 11).

All that business needs to do is ask the right question: "How do I transform my operations to *sustain* for the long term," think through its processes in detail, abandon the risky condition mind-set and tendency for risk transference, and engage in active experimentation and exploration toward "green" economic value exploration and creation; i.e., gains that are sustainable and do not produce waste(ful) inefficiencies. In adopting such an uncertain or ambiguous condition mind-set and working toward ensuring the safety of *our* children, as Craven (2009) so movingly asks, "What is the worst that can happen?"

At worst business would become sensitized to the wastage and inefficiencies in its present operations and learn what does not work. At best business just might trigger the required transformation culminating in the "both-and" development for Nature and Society that can sustain the children of all species and the ongoing prosperity of their generations.

References

Ackerman, R. W. 1975. *The social challenge to business.* Cambridge, MA: Harvard University Press.

Adam, B., U. Beck, and J. V. Loon. 2000. *The risk society and beyond: Critical issues for social theory.* Thousand Oaks, CA: SAGE.

Ahlstrom, J. 2010. Corporate response to CSO criticism: Decoupling the corporate responsibility discourse from business practice. *Corporate Social Responsibility and Environmental Management* 17(2): 70–80.

Alvarez, S. A., and J. B. Barney. 2005. How do entrepreneurs organize firms under conditions of uncertainty? *Journal of Management* 31(5): 776–793. doi:10.1177/0149206305279486

Anderson, R. C. 1998. *Mid-course correction: Toward a sustainable enterprise: The interface model.* Atlanta: Peregrinzilla Press.

Anderson, R. C., and R. A. White. 2011. *Business lessons from a radical industrialist.* New York: St. Martin's Griffin.

Aragón-Correa, J. A., and S. Sharma. 2003. A contingent resource-based view of proactive corporate environmental strategy. *Academy of Management Review* 28(1): 71–88.

Ardichvili, A., R. Cardozo, and S. Ray. 2003. A theory of entrepreneurial opportunity identification and development. *Journal of Business Venturing* 18(1): 105–123. doi:10.1016/S0883-9026(01)00068-4.

Ariely, D. 2009. A manager's guide to human irrationalities. *MIT Sloan Management Review* 50(2): 53–93.

———. 2010. *Predictably irrational: The hidden forces that shape our decisions.* New York: Harper Perennial.

Ash, M., and J. K. Boyce. 2011. Measuring corporate environmental justice performance. *Corporate Social Responsibility & Environmental Management* 18(2): 61–79. doi:10.1002/csr.238.

Bansal, P. 2002. The corporate challenges of sustainable development. *Academy of Management Executive* 16(2): 122–131.

Beck, U. 1992. *Risk society: Towards a new modernity.* Translated by M. Ritter. London: Sage.

———. 1996. World risk society as cosmopolitan society? Ecological questions in a framework of manufactured uncertainties. *Theory, Culture & Society* 13(4): 1.

———. 2009. *World at risk.* Cambridge, UK; Malden, MA: Polity.

———. 2010. Climate for change, or how to create a green modernity? *Theory, Culture and Society* 27(2/3): 254–266.

Benyus, J. M. 2002. *Biomimicry.* New York: Harper Perennial.

Boiral, O. 2006. Global warming: Should companies adopt a proactive strategy? *LongRange Planning* 39(3): 315–330. doi:10.1016/j.lrp.2006.07.002.

———. 2007. Corporate greening through ISO 14001: A rational myth? *Organization Science* 18(1): 127–146.

Boiral, O., M. Cayer, and C. Baron. 2009. The action logics of environmental leadership: A developmental perspective. *Journal of Business Ethics* 85(4): 479–499. doi:10.1007/s10551-008-9784-2.

Buehler, K., A. Freeman, and R. Hulme. 2008a. The new arsenal of risk management. *Harvard Business Review* 86(9): 92–100.

———. 2008b. Owning the right risks. *Harvard Business Review* 86(9): 102–110.

Busch, T., and V. H. Hoffmann. 2009. Ecology-driven real options: An investment framework for incorporating uncertainties in the context of the natural environment. *Journal of Business Ethics* 90(2): 295–310.

Craven, G. 2009. *What's the worst that could happen? A rational response to the climate change debate.* New York: Perigee.

Damasio, A. R. 1994. *Descartes' error: Emotion, reason, and the human brain.* New York: Putnam.

Delmas, M., V. H. Hoffmann, and M. Kuss. 2011. Under the tip of the iceberg: Absorptive capacity, environmental strategy, and competitive advantage. *Business & Society* 50(1): 116–154. doi:10.1177/0007650310394400.

Douglas, M., and A. B. Wildavsky. 1982. *Risk and culture: An essay on the selection of technical and environmental dangers.* Berkeley: University of California Press.

Dunn, S. 2002. Down to business on climate change. *Greener Management International* (39): 27–41.

Dyer, G. 2008. *Climate wars.* Toronto: Random House Canada.

Economist. 2011. A man-made world. *Economist* 399(8735), 81–83.

Epstein, S. 1994. Integration of the cognitive and the psychodynamic unconscious. *American Psychologist* 49(8): 709–724. doi:10.1037/0003-066X.49.8.709.

Farzad, R. 2011. The scrappiest car manufacturer in America. *BusinessWeek* (4232): 68–74.

Folta, T.B. 1998. Governance and uncertainty: The trade-off between administrative control and commitment. *Strategic Management Journal* 19(11): 1007–1028.

Ghemawat, P. 2003. Semiglobalization and international business strategy. *Journal of International Business Studies* 34(2), Focused Issue: The Future of Multinational Enterprise: 25 Years Later, 138–152.

Giddens, A. 1991. *Modernity and self-identity. Self and society in late modern age.* Cambridge: Polity Press.

Hawken, P. 1993. *The ecology of commerce: A declaration of sustainability*. New York: Harper Business.

Hoffman, A. J. 2005. Climate change strategy: The business logic behind voluntary greenhouse gas reductions. *California Management Review*, 47(3), 21–46.

———. 2011. Talking past each other? Cultural framing of skeptical and convinced logics in the climate change debate. *Organization & Environment* 24(1), 3–33. doi:10.1177/1086026611404336.

Hulme, M. 2009. *Why we disagree about climate change: Understanding controversy, inaction and opportunity*. New York: Cambridge University Press.

Inkpen, A. 1998. Learning, knowledge acquisition, and strategic alliances. *European Management Journal* 16(2), 223–229. doi: 10.1016/S0263-2373(97)00090-X.

International Energy Agency. 2010. *World energy outlook 2010*. Paris: International Energy Agency.

Islegen, O., and S. J. Reichelstein. 2009. The economics of carbon capture. *The Economists' Voice* 6(12) doi:10.2202/1553-3832.1685.

Japp, K. P., and I. Kusche. 2008. Systems theory and risk. In *Social theories of risk and uncertainty: An introduction*, ed. J. Zinn (p. 76). Malden, MA: Blackwell Pub.

Jordan, A. J., D. Huitema, H. van Asselt, T. Rayner, and F. Berkhout, eds. 2010. *Climate change policy in the European Union. Confronting the dilemmas of mitigation and adaptation?* Cambridge, UK: Cambridge University Press.

Kahan, D. M., P. Slovic, D. Braman, and J. Gastil. 2010. Fear of democracy: A cultural evaluation of Sunstein on risk. In *The feeling of risk. New perspectives on risk perception*, ed. P. Slovic, pp. 183–214. Washington, D.C.: Earthscan.

Kahneman, D., and A. Tversky. 1979. Prospect theory: An analysis of decision under risk. *Econometrica* 47(2): 263–291.

Kahneman, D., J. L. Knetsch, and R. H. Thaler. 1990. Experimental tests of the endowment effect and the Coase Theorem. *Journal of Political Economy* 98(6): 1352.

Knight, F. H. 1921. *Risk, uncertainty and profit*. New York: Houghton Mifflin Company.

Kolk, A., and J. Pinkse. 2004. Market strategies for climate change. *European Management Journal* 22(3): 304–314. doi:10.1016/j.emj.2004.04.011.

———. 2005. Business responses to climate change: Identifying emergent strategies. *California Management Review* 47(3): 6–20.

Langford, I. H. 2002. An existential approach to risk perception. *Risk Analysis: An International Journal* 22(1): 101–120.

Leiserowitz, A., E. Maibach, C. Roser-Renouf, and N. Smith. 2010. *Global warming's six Americas*. New Haven, CT: Yale University and George Mason University.

Lempert, R., J. Scheffran, and D. F. Sprinz. 2009. Methods for long-term environmental policy challenges. *Global Environmental Politics* 9(3): 106–133.

Loon, J. V. 2002. *Risk and technological culture: Towards a sociology of virulence*. New York: Routledge.

Luhmann, N. 1982. *The differentiation of society*. New York: Columbia University Press.

———. 1995. *Social systems*. Stanford, CA: Stanford University Press.

Lyng, S. 2008. Edgework, risk, and uncertainty. In *Social theories of risk and uncertainty*, ed. J. Zinn, pp. 106–137. Malden, MA: Blackwell Publishing Ltd.

McDonough, W. 2005. *Cradle to cradle*. Talk at the Technology, Entertainment and Design conference. Monterry, California. Full video and transcript of the talk available on the TED website at http://www.ted.com/talks/william_mcdonough_on_cradle_to_cradle_design.html (Accessed October 31, 2011).

McDonough, W., and M. Braungart. 2002. *Cradle to cradle: Remaking the way we make things*. New York: North Point Press.

McLeod, H., I. H. Langford, A. P. Jones, J. R. Stedman, R. J. Day, I. Lorenzoni, and I. J. Bateman. 2000. The relationship between socio-economic indicators and air pollution in England and Wales: Implications for environmental justice. *Regional Environmental Change* 1(2): 78–85.

Mendelsohn, R. 2000. Efficient adaptation to climate change, *Climate Change* 45(3/4): 583–600.

Miller, D., and I. L. Breton-Miller. 2005. *Managing for the long run*. Boston, MA: Harvard Business School Press.

Mishina, Y., B. J. Dykes, E. S. Block, and T. G. Pollock. 2010. Why 'good' firms do bad things: The effects of high aspirations, high expectations, and prominence on the incidence of corporate illegality. *Academy of Management Journal* 53(4), 701–722.

Nitin, D., R. Foster, and J. Medalye. 2009. A systematic review of the literature on business adaptation to climate change. *Network for Business Sustainability*. http://nbs.net/knowledge/climate-change/systematic-review/ (Accessed July 31, 2011).

NRC. 2010a. Adapting to the impacts of climate change. *America's Climate Choices: Panel on Adapting to the Impacts of Climate Change; National Research Council* 292.

———. 2010b. America's climate choices. *America's Climate Choices: Committee on America's Climate Choices; National Research Council* 144.

———. 2010c. Informing an effective response to climate change. *America's Climate Choices: Panel on Informing Effective Decisions and Actions Related to Climate Change; National Research Council* 348.

———. 2010d. Limiting the magnitude of future climate change. *America's Climate Choices: Panel on Limiting the Magnitude of Future Climate Change; National Research Council* 276.

O'Malley, P. ed. 2008. *Governmentality and risk*. Malden, MA: Blackwell Pub.

Parsons, T. 1980. Health, uncertainty and the action structure. In *Uncertainty. Behavioural and Social Dimensions*, ed. S. Fiddle. New York: Praeger.

Pinkse, J., and A. Kolk. 2010. Challenges and trade-offs in corporate innovation for climate change. *Business Strategy & the Environment (John Wiley & Sons, Inc)* 19(4): 261–272.

Quinn, D. 1992. *Ishmael: An adventure of the mind and spirit*. New York: Bantam/Turner.

Sethi, S. P. 1979. A conceptual framework for environmental analysis of social issues and evaluation of business response patterns. *Academy of Management Review* 4(1), 63–74.

Slovic, P. 2010. *The feeling of Risk: New perspectives on risk perception.* Washington, D.C.: Earthscan, in association with the International Institute for Environment and Development.

Slovic, S., and P. Slovic. 2010. Numbers and nerves: Towards an affective apprehension of environmental risk. In *The feeling of risk: new perspectives on risk perception,* ed. P. Slovic, pp. 79–82. Washington, D.C.: Earthscan.

Smith, C. W., and R. M. Stulz. 1985.The determinants of firms' hedging policies. *Journal of Financial and Quantitative Analysis* 20(4): 391–405.

Steffen, A. 2006. *Worldchanging: A user's guide for the 21st century.* New York: Abrams.

Sunstein, C. R. 2005. *Laws of fear: Beyond the precautionary principle.* New York: Cambridge University Press.

Tainter, J. A. 1988. *The collapse of complex societies.* New York: Cambridge University Press.

Thompson, J. D. 1967. *Organizations in action; social science bases of administrative theory.* New York: McGraw-Hill.

Thompson, L. L., J. Wang, and B. C. Gunia. 2010. Negotiation. *Annual Review of Psychology* 61, 491–515.

Thornhill, S., and R. Amit. 2003. *Learning from failure: Organizational mortality and the resource-based view.* Ottawa: Statistics Canada, Micro-Economic Analysis Division.

Tversky, A., and D. Kahneman. 1974. Judgment under uncertainty: Heuristics and biases. *Science* 185(4157): 1124–1131.

———. 1981. The framing of decisions and the psychology of choice. *Science* 211(4481): 453–458.

———. 2004. Loss aversion in riskless choice: A reference-dependent model. In *Preference, belief, and similarity: Selected writings by Amos Tversky,* ed. E. Shafir and E. Shafir, pp. 895–915. Cambridge, MA: MIT Press.

Wade-Benzoni, K., A. J. Hoffman, L. L. Thompson, D. A. Moore, J. J. Gillespie, and M. H. Bazerman. 2002. Barriers to resolution in ideologically based negotiations: The role of values and institutions. *Academy of Management Review* 27(1): 41–57.

Zinn, J. ed. 2008. *Social theories of risk and uncertainty: An introduction.* Malden, MA: Blackwell Pub.

PART II

Key Players Managing the Risks of Climate Change

CHAPTER 3

The Primary Insurance Industry's Role in Managing Climate Change Risks and Opportunities

Lára Jóhannsdóttir, James Wallace, and Aled Jones

Climate change presents a number of risks to economies and societies. These risks include physical, market, policy, and security risks. Some of these risks are more immediate than others, but all need careful management over time. The insurance sector has a critical role to play in tackling climate risks through the provision of risk management expertise and the provision of risk transfer mechanisms. The insurance sector is also one of the business sectors highly exposed to climate change risks because of the likely increase in frequency and severity of weather events and resultant socioeconomic changes. At the same time, climate change offers opportunities for the insurance industry, but to seize these opportunities, each insurer will need to focus on certain types of risks and products that are within its risk appetite. This chapter addresses three questions: *What does climate risk mean for insurers? How can climate risk be removed and/or reduced?* And *what are the climate opportunities for insurers?*

What Does Climate Risk Mean for Insurers?

Some physical, market, policy, and security risks created by climate change are more immediate than others but all need careful management over time. Bold leadership on climate change may mitigate some of these risks; however, it is evident that global society will not respond with the urgency and scale required to eliminate all of these risks. Therefore, new and emergent risks will become increasingly important for all parts of global and national economies—and the insurance sector in particular will need to understand and manage these risks if society is to function well (and remain insurable).

The insurance sector has a critical role to play in tackling these risks. As stated by ClimateWise (2009), insurers fulfill two key roles in managing these risks—the provision of risk management expertise and the provision of risk transfer mechanisms (the core business of insurance).

Vulnerable Societies and Impacts on Insurers

Climate change represents a significant threat to all societies although the extent of its impacts remains uncertain. While it is "virtually certain" that increased greenhouse gases in the atmosphere will continue to cause a significant rise in the global average temperature and significant, perhaps dramatic, changes in weather patterns, what such changes will mean at the local level remains a topic of intense scientific research (IPCC, 2007b). Some areas of this uncertainty are more amenable to sophisticated modeling than others and in those areas quantifiable measures of risk can be developed. Nevertheless, the twin challenges of uncertainty and risk mean the insurance sector has an important role in managing the climate change future and will require the industry to use both its risk management expertise to model uncertainty more effectively and its risk transfer mechanisms to manage the known risks.

The key uncertainty in modeling climate change is societal response to this challenge. This uncertainty is very difficult to measure and quantify and thus the insurance sector, alongside other sectors, needs to develop "best guess" models for future scenarios under which risk is quantifiable and underwriting of the risk is possible. In some current scenarios, where response to climate change is delayed and does not meet the size and scope of the challenge, it is increasingly clear that the risks associated with climate change will not be insurable—they will occur too frequently and/or with too great an impact to make insurance a viable risk management tool (ClimateWise, 2009).

Therefore, when looking at risk management in the future, we need to understand the natural, institutional, and socioeconomic vulnerability that forms the backdrop of any scenario. The insurance sector is starting to respond to some of these challenges by collaborating on policy, research, and data. One of these collaborative initiatives is ClimateWise, a group of over 40 global insurers that launched the ClimateWise principles in 2007 in collaboration with the University of Cambridge Programme for Sustainability Leadership. Another of these initiatives is the Geneva Association, whose membership comprises a statutory maximum of 90 Chief Executive Officers (CEOs) from the world's top (re)insurance companies. The Geneva Association is the leading international insurance think tank for strategically important insurance and risk management issues. The United Nations Environment Programme (UNEP) Finance Initiative is another prime example.

Assuming that society does respond in a meaningful way to climate change and allows insurance to remain a useful risk management tool, the main risks for the insurance sector will be physical, policy, market, and security risks. While only the first of these is related to the physics of climate change, all are influenced by the socioeconomic aspects of climate change.

Physical Risk

Specific physical impacts from climate change are difficult to verify because a particular weather event cannot currently be directly associated with climate change. Climate change alters the probability distribution of certain events, but changes in the probability distribution also alter the risk profile and therefore have a direct impact on the viability of insurance for certain events, whether they be tropical thunderstorms in Asia or droughts in Africa. At present, there is little evidence that physical impacts are currently changing the risk profile of certain events. Although researchers such as Munich Re have shown that over recent years there has been a significant increase in financial exposure to weather-related events, the cause of this increase may not arise solely from climate change, but is more likely to be related to the increased value of properties and infrastructure located in vulnerable regions, such as coastlines in North America (Munich Re, 2011). In addition, the lack of comprehensive historical data for financial losses due to weather events at a global level limits the conclusions that can be drawn. However, 2010 did yield the second highest (after 2007) number of loss-related weather catastrophes since 1980 when appropriate data collection began (Munich Re, 2011).

Over time the physical impacts of climate change will differ dramatically and a key issue for the insurance sector will be finding ways to incorporate changing risk profiles into the real risk probabilities that are used in pricing decisions and in developing products to manage new risks. Initially, weather-related impacts from climate change will be felt most severely in those sectors and industries that are heavily reliant on weather, such as agriculture, tourism, utilities, property, and health, as well as those industries whose supply chain or operations are based in vulnerable regions (Reiman, 2007).

Table 3.1 summarizes some of the expected physical impacts from climate change within this century that need to be incorporated and modeled in insurance product development.

Table 3.1 Physical climate risk phenomena as indicated in Intergovernmental Panel on Climate Change 4th Assessment Report

Phenomenon and direction of trend	*Likelihood of occurrence in twenty-first century*[a]	*Human Health*	*Industry, settlement and society*
Over most land areas, warmer and fewer cold days and nights, warmer and more frequent hot days and nights	Virtually certain	Reduced human mortality from decreased cold exposure	Reduced energy demand for heating; increased demand for cooling; declining air quality in cities; reduced disruption to transport due to snow, ice; effects on winter tourism
Warm spells/heat waves. Frequency increases over most land areas	Very likely	Increase risk of heat-related mortality, especially for the elderly, chronically sick, very young and socially isolated	Reduction in quality of life for people in warm areas without appropriate housing; impacts on the elderly, very young and poor
Heavy precipitation events. Frequency increases over most areas	Very likely	Increased risks of deaths, injuries and infectious, respiratory and skin diseases	Disruption of settlements, commerce, transport, and societies due to flooding; pressures on urban and rural infrastructure; loss of property
Increased incidence of extreme high sea level (excludes tsunamis)	Likely	Increased risk of deaths and injuries by drowning in floods; migration related health effects	Costs of coastal protection versus costs of land-use relocation; potential for movement of populations and infrastructure

[a] Based on IPCC projects: Virtually certain (>99% probability of occurrence), Very likely (90–99% probability of occurrence), Likely (66–90% probability of occurrence).
Source: IPCC 2007a.

Policy Risk

The uncertainty created by the physical risks arising from climate change is compounded by uncertainty about political responses to the resulting challenges of climate change. These uncertainties, such as changes in taxation and regulation, can create additional costs, shifts in consumption, and changes to the market that in turn create different valuations for and impacts on insured assets and processes. This policy uncertainty is much more difficult to model than physical risk. However, it is much more likely to have a significant impact in the short term. Therefore, understanding the likely outcome of policy negotiations and developments is vital for enabling a thriving insurance market. Mitigation technologies in particular, such as renewable energy, very often rely on stable and long-term policies to ensure growth in the market. Adaptation approaches, such as flood plain management infrastructure, are often heavily reliant on public finance and investment. Any changes in future policy, such as planning regulations, can have a significant impact on the risk profile for assets and infrastructure in flood-prone zones.

Market Risk

There is little current evidence that market drivers are enabling solutions to climate change to be deployed at scale. The overarching situation is one of "business as usual" (International Energy Agency (IEA), 2010). Different industry sectors will be exposed to different market drivers over time, whether they are consumer-led, policy-led, or technology-led. The different market drivers will have more impact in certain sectors than in others. For example, rapid developments in technology in a sector will lead to changes in products that in turn will change risk exposures of companies in that sector (The Carbon Trust, 2008). At present it seems unlikely that any sector will experience a net reduction in risk profile as a result of climate change, therefore there is likely to be a net increase in risks across the market (Reiman, 2007).

Security Risk

Increasing stress on food, water, and energy supplies could result in worsening regional, national, and international relations. Such worsening relations could lead to an increase in local disruption, which might threaten supply chains and infrastructure. It has been suggested that companies will need to plan for increasing uncertainty of supply in the future as a result of climate-change-induced security issues (Lloyd's 360 Risk Insight, 2009).

Climate Risks and the Business of Insurance

As noted above, there are many risks associated with climate change. These risks can impact the non-life insurance and life insurance lines of business, the investments of insurance companies, insured clients, and even existing and new insurance products.

Non-Life Insurance

Property

Property will be significantly impacted by climate change over the long term and perhaps medium or even short term. Both extreme weather events and long-term temperature changes will impact the risk profiles of infrastructure. Examples include flooding, long-term warming resulting in drier soils leading to subsidence, and property in coastal areas becoming uninsurable as sea levels rise over the next century.

Liability

Under changing legislative environments, organizations will increasingly be exposed to the threat of lawsuits. As knowledge of climate change impacts increases, the ability of companies to foresee particular events and the likelihood of being held partly responsible for those events will become even more pressing issues. For example, should a property developer have built homes in a flood plain? Should the planning authority have allowed the property to be built? Should a company drill for fossil fuels when the use of those fuels will increase greenhouse gas (GHG) emissions?

Business Interruption

Business interruption insurance will increasingly be impacted by climate change. Sectors that are heavily reliant on the weather, or those with supply chains and operations in vulnerable regions, will see larger and longer interruptions to business as the physical impacts of climate change increase.

Other Non-Life Products

There are also additional indirect risks from climate-change-driven changes in technologies and consumption patterns, particularly in transportation and agriculture. These changes may impact insurance products such as car insurance or crop insurance in unpredictable ways. New risks will emerge as technology solutions are deployed with associated insurance products. For example, political insurance against changing subsidies for renewable energy generation will obviously depend heavily on a good understanding of the likelihood of changing governmental policies.

Life Insurance

Health

The impact of global warming and extreme weather events on individuals has already been felt, and is discussed in *Changing Planet, Changing Health* by Paul Epstein and Dan Ferber (2011). The heat wave in Europe in 2003 caused 52,452 deaths (Earth Policy Institute, 2006), while the floods in Pakistan in 2010 caused 1,802 deaths and impacted almost 1.4 million people (National Disaster Management Authority, 2010). It is likely that these extreme weather events will become more frequent and, unless steps are taken to manage the impact of heat waves, there will be increasing numbers of deaths or hospitalizations. On the other hand, milder winters may result in fewer cold-related deaths. Understanding this changing risk profile will be critical to maintaining a life insurance policy portfolio.

The long-term impact of climate change will also have an effect on health—both through the availability of food and water as well as the impact on living conditions in cities and built-up areas.

The impact of extreme events such as flooding on mental health should not be underestimated. Several reports indicate that flooding has a significant effect on mental health and depression (Few, Ahern, Matthies, and et al., 2004). Those affected by these events may not immediately show symptoms and therefore long-term studies are required to understand the true extent of these impacts.

However, the most vulnerable populations for health impacts (older and younger people) may not be the most insured—in part because they are less likely to be insured through work-based insurance products. Therefore, the impact of climate change on life insurance may be mitigated slightly in the short term. On the other hand, there is a clear political drive to make access to health care and health-care services more equitable, which in turn would increase the insurance sector exposure to the health impacts of climate change (Baum and Fisher, 2010).

Investments

All of the risks identified in this section can influence the valuation of investments made by the insurance sector. However, with a few notable exceptions, insurance companies are not currently incorporating climate risk into their investment decisions. It is believed that this situation results primarily from the companies' views on regulatory weakness; both regulation to drive the necessary change to deal with climate change as well as the potential for concessions to mitigate value-at-risk (The Carbon Trust, 2008). With increasing regulation in this sector, such as the European Union Emissions Trading Scheme (EU ETS), it will not be long before companies will see climate risk's impact on the valuations of their investments. In addition while it has been very difficult to get legislation within the United States associated with climate change, the new Securities and Exchange Commission Guidance on

Disclosure Related to Business or Legal Developments Regarding Climate Change (2010) offers some evidence that movements in this direction are inevitable.

With fixed income assets there is also increasing evidence that certain bonds have hidden climate risks that have not been taken into account fully in their pricing. For example, a recent study (Leurig and PricewaterhouseCoopers LLP, 2010) examines the hidden risks in utility bonds in North America due to changing water availability in the regions where these utilities operate. These hidden risks in fixed income assets could become more pronounced if action to reduce the impact of climate change is not significant. For example, government bonds could be impacted if climate risk in certain regions starts to have a significant impact on the Gross Domestic Product (GDP) or the social cohesion of those countries.

Indirect Impacts (via Clients)

Indirect impacts are varied and are difficult to quantify. All of the risks identified above may have a significant impact on clients, individuals, or organizations, even if the impacts do not fall upon insured risks. The impacts could start a chain of events that then damage the insured parts of a client's operations. For example, even if liability insurance associated with climate change is specifically excluded from coverage and an organization (or individual) is found liable for a particular impact, the client may go out of business or an entire sector of the economy could even fall into decline.

Radical Greening as a Threat (and Opportunity) for Insurance Products

A final issue to be considered by the insurance sector is the trend toward more "green" insurance products and services. While there is little evidence that increasing consumer demand for green products is at a scale that would change insurance markets, consumer awareness is increasing. If over time a real consumer trend is observed—"radical greening"—both for insurance products and also for the products of clients, then the insurance sector will have to react to this demand quickly (Ernst & Young and Oxford Analytica, 2009). If large impacts from weather-related events (such as a major hurricane or drought) drive a more rapid move toward "green" insurance products, then the ability of the insurance sector to respond rapidly will be vital to preserving markets and profitability.

How Can Climate Risk be Dealt with and/or Reduced?

On global, regional, and local levels, climate risk must be reduced through the implementation of mitigation strategies to reduce the impact of climate change and through adaptation to manage the impact of climate change. However, even very effective mitigation and adaptation strategies will still yield residual risk that can be spread through risk-transfer mechanisms such as insurance. The discussion in this section is not intended to be exhaustive, but merely to give an overview of how insurers can deal with this residual climate risk.

The insurance sector has been identified as one of the sectors highly exposed to climate change because of more frequent and severe weather events, global warming, and socioeconomic changes. Socioeconomic changes include such phenomena as a growing world population and urbanization that concentrate assets and individuals in at-risk areas such as coastal areas. As the previous section illustrates, climate risk affects different lines of the insurance business in different ways. The new challenges created by changing climate and weather patterns are expected to intensify in the future (IPCC, 2007a).

A recent global survey of 60 financial institutions reveals that both historical weather data and climate predictions are needed if rising climatic risks affecting the insurance portfolios are to be managed (Clements-Hunt et al., 2011). Insurers must find ways to develop relevant information on the physical and economic consequences of climate change so they can take length of insurance contracts and location of their client's assets and operations into consideration in developing and pricing products and so they can offer risk advice to their clients. The study also noted that the demand for additional risk-transfer capacity is increasing and will continue to do so, that insurance products are being modified and will have to be revised, and that insurers are in the phase of developing new products, or have further plans for doing so. It emphasizes that insurers should look beyond the lifetime of insurance contracts: while their actual risk exposure may be limited by the specific contracts in place, there is a risk that future markets could become uninsurable and therefore market opportunities could diminish.

Ways of Managing Risk

In classic risk management literature there are four main strategies for dealing with risks: avoid, accept (retain), reduce, or share (transfer) the risk (see fig. 3.1) (Gibbs and DeLoach, 2006).

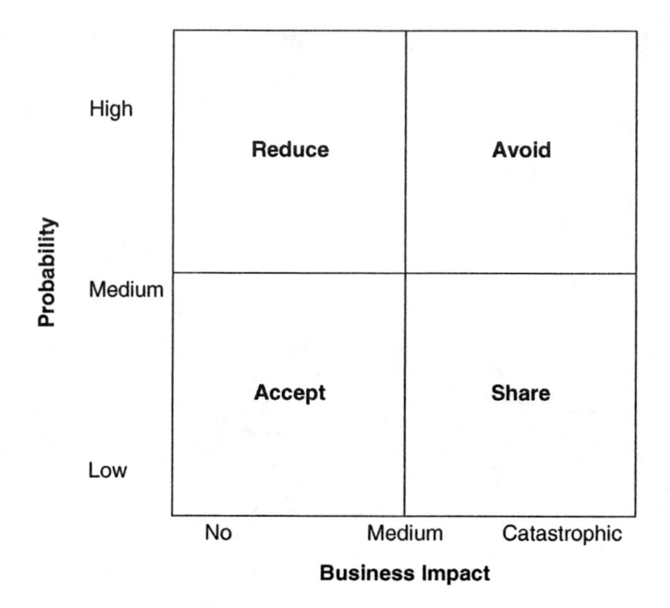

Figure 3.1 Four main strategies for dealing with risk.

On a more aggressive scale, risk can be exploited (DeLoach, 2000; Lessard and Lucea, 2006) or in some cases ignored (Tomlin, 2006). In addressing the question of how insurance companies can remove or reduce climate change risks, we discuss ways primary insurers can protect their own business, not how they can reduce the risk of their clients—even though in cases where insurers can influence the risk exposure of their clients it will transfer positively into their own risk portfolios.

Avoid Risk

When there is a high probability of catastrophic events with negative business impacts the risk should be avoided. Limiting the availability of insurance products would be one approach to avoiding the risk, but insurers can also signal their risk avoidance through price policies and insurance terms and conditions (Mills, 2009). When lack of sufficient data to evaluate and price the risk results in insurers being reluctant to take on the risk, they can avoid the risk by "pricing themselves out of particular markets" with high premiums and strict terms and conditions. Insurers can also place new exclusions on or choose not to renew already existing insurance contracts (Mills, 2005; Perroy, 2005). On limited scales such actions are already being taken. In the UK and Germany insurers have already started to exclude from insurance the properties most exposed to climate change events (Perroy, 2005). These exclusions are occurring because the governments have failed to invest in flood defenses (Brown, 2007), so future flood risk remains high and risk of damaged properties remains strong. In the US states such as Florida, some homeowners have already lost coverage on their properties (Mills, November 2007; The Geneva Association, 2009). In these cases, insurers are segregating higher risk from lower risk and are avoiding the higher risk because insuring such risk can have a negative impact on their business performance and perhaps even threaten their solvency. The side-effect to this strategy is that insurers can become scapegoats because their actions shift the burden to the state, which is a situation that is not politically, socially (Mills, 2003), or financially desirable. As a consequence, insurers might lose bargaining power with governments (The Geneva Association, 2009) and their credibility as risk experts.

Reduce Risk

In cases of high probability of climate risk, with little to medium business impact, insurers can take on the risks but the strategy would be to reduce the severity of losses or the likelihood of losses occurring. In such cases insurers can reduce the variance of the risk by spreading it over time and among different policyholders, thereby limiting their exposure to losses. Analyzing loss trends, promoting loss-minimizing or loss-preventing initiatives, promoting incentives for loss reductions, sharing risk management expertise, or raising climate awareness among stakeholders can be used to reduce the climate risk (Climate Change Policy & Practice, 2010). From environmental and climate perspectives, loss prevented is always better than loss compensated. Insurers are also increasing premiums and deductibles, tightening insurance terms by lowering limits on protection, placing caps on the concentration of claims (Mills, 2005; Perroy, 2005), and raising standards or placing demands on suppliers (Jóhannsdóttir, 2009). Under the risk-reduction strategy, insurers can provide guidance to their clients, influencing their behavior and thereby reducing the risk. In cases of risk avoidance and risk reduction where insurers start to withdraw from markets, raise premiums, use deductibles, or exclude weather-related risks, the burden shifts over to the public, the business environment, and governments, which then "become insurers of last resort" (Mills, Roth, and Lecomte, 2005, p. 2).

Accept (Retain) Risk

Although it might be logical to avoid or reduce climate risk entirely, insurers might lose out on potential gains or profits they can obtain by accepting the risk. If the probability of specific climate risk is low to medium, with little to medium business impacts, insurers should retain the risk. In many cases insurers buy reinsurance treaties for the climate risk; however, there remains self-risk (risk that is retained by the insurance company). For a risk that exceeds acceptable levels of self-risk there are two main reinsurance forms available: treaties and facultative contracts.

Reinsurance treaties are used to reinsure particular portfolios against certain risks. The two main types of such treaties are the excess of loss (stop-loss) and the quote share treaties. In excess of loss treaties, reinsurers assume upper layers of risk, paying out after a given threshold is reached (Charpentier, 2008), while insurers retain the remaining risk. In typical excess of loss treaties there is a single policy risk. In "cat" excess of loss treaties the payment, after the self-risk is reached, is based on a loss caused by a single catastrophic event such as flood or a hurricane. In quote share treaties the risk shared among insurers and reinsurers is proportional. If insurer and reinsurers have reached an agreement of 20 percent quota share the insurer will transfer 20 percent of the premiums to the reinsurer, which will accept 20 percent of the risk written by the insurer within the category of the risk (e.g., properties) defined in the treaty. This arrangement means that 20 percent of premiums and liability is transferred from the insurer to the reinsurer, who has to pay out 20 percent of losses sustained by the insurer, whether it is partial or total loss.

Facultative reinsurance treaties provide cover of specific individual risks that are not covered in other reinsurers treaties and have to be individually evaluated and insured, due to the size of the risk or because the risk is unusual. This type of treaty could be for an oil tanker or an oil rig, for example. Of course, reinsurers are free to accept or deny taking on the risk depending on how they evaluate it. It is therefore up to primary insurers to assess whether they are willing to underwrite the risk as a self-risk or reject it entirely.

Share or Transfer Risk

In general, primary insurers have the role of carrying their clients' risk, while reinsurers cover insurance companies for higher concentrations of risks. In the case of low to medium probability of catastrophic events with negative business impacts, insurers can take on the risk by sharing or transferring it. They can share the risk among a group of policyholders or they can transfer their risk by buying reinsurance coverage as outlined in the last section. Reinsurers are able to spread the risk on a more global scale than insurers. However, there are also limits to what the reinsurance sector can absorb when it comes to climate risk. As Conning & Company (1994) point out, Hurricane Andrew had a significant impact on the insurance industry and nine insurance companies in Florida alone became insolvent (Charpentier, 2008).

Insurers are utilizing additional options for transferring risk. Climate risk can be transferred through capital markets via Insurance Linked Securities (ILS) such as catastrophic (CAT) bonds driven by loss events. CAT bonds are risk-related securities whose return depends on the occurrence of a specified catastrophic event. CAT bonds are issued by sponsors, such as insurers or governments. There is a high risk / low probability embodied in the bonds and therefore they offer high interest rates, making the bonds attractive to investors. In the case when certain events trigger the CAT bonds' agreements, the investors lose their investments. "Reinsurance sidecars" are instruments used with CAT bonds where private investors, such as hedge funds, are able to take on the risk underwritten by (re)insurers with the potential of earning return on the risk they take. Weather derivatives are yet another type of risk spreading instruments that are used to insure against abnormal changes in temperature, for example, or in the event of natural disasters.

Exploit Risk

Insurers also have the option of exploiting opportunities (De Loach, 2000; Lessard and Lucea, 2006) stemming from climate risk and thereby gaining competitive advantages if the outcome is positive. Such opportunities could occur, for example, in cases of insuring renewable energy technologies.

Ignore Risk

Insurers might ignore or overlook the climate risk because they do not see any climate-related trends in their claims statistics (Tomlin, 2006). In other cases they might be inactive and thereby accept the risk by default or they might lack information to act on the local impact of climate change and choose to take no action. Such decisions might, of course, later lead to serious losses in the case of catastrophic events. Even though the number of primary insurers that are proactively tackling the climate issue is rising (Mills, 2009), wait-and-see climate strategies still exist among insurers (Mills, 2003) because the insurance sector still appears to be reactive in general (The Geneva Association, 2009). One reason why insurers might ignore or overlook climate risk is because of the conservative backward-looking perspectives and long-established expertise of insurers, where risk is identified and quantified statistically and priced according to historical data. A typical insurer does not have the expertise to include a vision of a sustainable low-carbon economy and the necessary tools to model future risks (Stahel, 2010). Climate change as a business driver for primary insurers has still not been widely recognized.

Climate Actions of Insurers

Adaptation

Since anthropogenic climate change caused by industrialization and land-use changes became an emerging issue globally, the main focus of world leaders, the scientific community, and environmentalists has been on reducing GHG emissions. The aim of the Kyoto Protocal, for example, was to set binding emission targets, and focus on emission trading, clean development mechanism (CDM), and joint implementation (JI). Taking actions to adapt to the consequences was neglected to a great extent because it was considered "politically incorrect" to take actions that implied defeat on the issue of climate changes. Collective actions are needed, but governments from around the world have not yet been able to reach political agreement on reducing global GHG emissions. Under this "business as usual" situation, GHG emissions will continue to grow over the following few decades, with some irreversible negative impacts (IPCC, 2007a). It is those impacts that insurers will feel for decades into the future (The Geneva Association, 2009). Therefore, it is of vital importance for the insurance sector to promote adaptation actions as a means to reduce climate risk and to ensure the availability and affordability of insurance products. In this respect the core of the insurance business, risk management and risk expertise, is a valuable contribution to the resolution of the climate issue.

The main adaptation themes identified relate to storms, flooding, agriculture (additional yield in some places, crop failure in others), water availability, heat, and health issues (The Geneva Association, 2009). In many cases, insurers can contribute to the solutions of those issues by offering new insurance products, stimulating behavioral changes, and through claims handling processes and risk expertise. While assessment of regional differences in risk and land-use planning can be used to reduce insurers' exposure to climate events, such assessments can also help clients take actions that reduce their exposure to climate risk and encourage them to undertake loss prevention activities. Public-private partnership in some infrastructure projects and development of new building standards can improve resilience to catastrophic incidents, such as hurricanes and floods.

Mitigation

As predominantly financial institutions offering intangible services in the form of "peace of mind," insurers have a "low carbon footprint" from their direct operations. Tracking and reducing a company's direct carbon footprint is a common and simple starting point for many insurers. Energy review, energy efficiency in buildings, and business travel are among the initiatives insurers focus on. However, when claims occur the intangible service becomes tangible as the claims have to be handled (Meyricke and ClimateWise Sustainable Claims Steering Group, 2010). Handling claims does have carbon footprint impacts through wasteful use of new resources, transportation, et cetera. Handling claims and developing innovative new services are fields where insurers can have the most impact on mitigation and moving toward low-carbon economies. The main themes for reducing GHG emissions are usage of solar energy, biomass and biofuels, wind farms, geothermal power, hydropower, nuclear power, carbon capture and sequestration, forestry, energy efficiency, transport sector efficiency, buildings, efficiency of industries, and geoengineering (The Geneva Association, 2009). In general, the insurance sector is in the position of promoting development in these areas by assessing and pricing new products and services that speed up the deployment and use of low-carbon technologies.

Developing insurance products focusing on reducing GHG emissions for buildings and transportation are among the initiatives insurers can take (IPCC, 2007b). Pay-as-you-drive (PAYD) auto insurance products based on per mile driven, instead of lump sum payments, are estimated to reduce driving by 8 percent, oil consumption by 4 percent, CO_2 emissions by 2 percent, and to reduce driving-related accidents (Bordoff and Noel, 2008). Claims processes that encourage or specify use of second-hand spare parts in car repairs save huge amounts of carbon emissions, in addition to being economically attractive for insurers (Stahel, 2010). Incentives for buying fuel-efficient motor vehicles or vehicles using alternative energy sources are also mitigation solutions. Restoring or rebuilding houses in more energy-efficient ways, or promoting passive energy designs, or even "nega-watt" buildings, which produce more energy than they use, all have mitigating effects in addition to having positive impacts on public safety and financial returns (Stahel, 2010).

The Interlink Between Adaptation and Mitigation

According to the IPCC there can be synergies and trade-offs between adaptation and mitigation (IPCC, 2007). For the insurance sector, synergies between adaptation and mitigation are often possible and feasible, such as in the building and the energy sectors, agriculture, forestry, and land use (Mills, 2006). Among the examples Mills offers are energy demand and supply where solutions can reduce energy consumption and related emissions, while offering adaptation benefits on grid reliability and backup generation requirements during outages. Grid reliability and backup facilities have positive impacts on business interruption for insurers. Improved forest management will reduce the potential of wildfires and carbon-related emissions, with the insurance benefits related to property, health, and life. New building codes can lead to positive mitigation and adaptation results simultaneously.

Adaptation in the Developing Countries

Lower-income populations are particularly vulnerable to climate risk, even though they are the least responsible for the changing climate. Climate risk hampers development in lower-income countries, where even minor economic losses can have long-term negative impact on economic development and human health (The Geneva Association, 2009; UNEP, 2010). In Africa areas of specific concern are "water resources, agriculture, health, ecosystems and biodiversity, forestry and costal zones" (Africa Partnership Forum, 2007, p. 8).

While insurance mechanisms, and other risk-transfer mechanisms, are well established in the developed countries, they are not well established in developing and emerging economies where risk-transfer mechanisms for natural disasters are underdeveloped and still limited if they exist at all (Arnold, 2008). There is a huge penetration gap in insurance availability between the developed world, developing world, and emerging markets. Climate change consequences in lower-income countries can slow down economic activities and interrupt supply chains worldwide as explained earlier. Insurance leaders within the UN Environment Programme Finance Initiative (UNEP FI), ClimateWise, The Geneva Association, and the Munich Climate Insurance Initiative (MCII) are now urging governments globally to utilize their expertise and risk management knowledge to enhance resilience, particularly in the developing world (Climate Change Policy & Practice, 2010; UNEP, 2010).

Insurance solutions are being explored in the poorer parts of the world on micro-, meso-, and macro-levels (Mechler, 2008). Risk is being pooled among countries using index-based solutions, public-private partnerships are taking place, and micro-insurance solutions are offered to farmers and poor households (The Geneva Association, 2009). Index-insurance solutions are being used to unlock development potential on individual and community scales, as a mechanism for disaster relief on national and multinational scales, and in the context of climate change adaptation to address problems facing the poor and the vulnerable (Hellmuth, Osgood, Hess, et al., 2009). Index insurance is used in cases such as loss of livestock, loss of crops due to droughts, and losses due to weather events such as hurricanes. Humidity, precipitation, temperature, or yields of harvests are currently used as indexes. These approaches have been used for flood insurance in the Mekong Delta in Vietnam, for livestock insurance in Mongolia, and for catastrophic risk insurance in the Caribbean. Benefits of such products are lower transaction costs, less moral hazard, and less adverse selection relative to conventional insurance. Reinsurers such as Swiss Re have, together with Oxfam America and local partners, been exploring weather insurance products for the poor in Ethiopia through labor-for-premium insurance schemes (Swiss Re, September 2010).

What are the Climate Opportunities for Insurers?

Opportunities for the insurance industry can be viewed from a number of perspectives. Many perceive the opportunities in selling green personal products and carbon-related innovations. But there are many other opportunities for insurers and they can have a beneficial impact on climate change. Each insurance company will consider climate change opportunities depending upon a number of criteria that are explored below. The discussion will not present an exhaustive list but will highlight some of the considerations insurers take into account.

While insurance plays an important risk transfer and management role, it is not a solution for all aspects of climate change. There is simply not enough capital within the insurance system to cover all risks and not everything is insurable. If a catastrophic climatic scenario can be valued in terms of financial impact, consideration needs to be given as to whether a policy can be written and if there is a customer willing to pay the premium.

The Language of Climate Change

The term climate change can be a polarizing term within the industry. The industry has been in existence for over 300 years and has seen a wealth of different risks, but perhaps none of this potential magnitude. An early challenge for insurers involves identifying what specific risks climate change presents to customers beyond the normal cover of weather-related events.

One of the difficulties climate change presents is the difficulty of determining its impact on consumers, whether the impact falls on their home, business, pet, car, or something more exotic. In attempting to disaggregate the risks of climate change, those risks can often be broken down into some of the traditional insurance risks such as wind damage, flooding, subsidence, or other forms of extreme weather damage that could lead to a claim. These types of risks are generally considered as standard by the insurer through careful analysis of the risk profile of the customer.

While the intensity and frequency of these risks may be changing with the onset of climate change, these changes may not yet be indicated by the insurers' own historical claims data. While some sources might consider this perspective to be an overly backwards-focused and conservative view (Stahel, 2010), the degree of uncertainty of these risks means insurers may feel they cannot pass their estimates of the true costs onto the customer with a clear conscience.

Climate change is best considered in the industry when it is discussed in terms of specific risks that customers can be directly exposed to within their policies. This approach is something that the industry understands and is fully prepared to step up to the challenges it presents.

Considering the Risk Appetite

Climate change is not a new issue for the insurance industry. For the majority of organizations it has been on emerging risk registers for over ten years. Each insurer will focus on certain types of risks and products, which will be deemed either within or outside their risk appetite. Where the risk is novel or new, the potential liability is often unknown, so a greater risk appetite is needed if the insurer is to take on the risk. This route may be appropriate if the insurer feels it occurs in a growth area where a first mover approach will establish market-leading expertise. Many of the more exotic climate change catastrophe bonds or the underwriting of carbon offsets fall into this category and at present can be considered to be niche markets.

Each insurer is also aware of the potential long-tail liability, i.e., potential claims arising long after the initial writing of the policy. Asbestos is a key example of such a situation, having initially been considered a benign material with only positive applications. Because of such possibilities, the terms of the policies need to be written with great clarity to protect the balance sheet of the insurer, provide customers with specific understanding of their cover, and protect the interests of shareholders.

The new Solvency II EU Legislation is due to come into force to ensure insurers maintain enough capital to cover claims. This legislation may encourage more conservative approaches to the risks being underwritten or investments being selected. Insurers are therefore less likely to be involved in new prototype clean technology or niche, high-risk, markets like carbon insurance schemes.

The Role of Legislation

Insurance can play an important role in providing early warning signals to governments and consumers. Many insurers play a proactive and responsible role in raising issues with government to ensure affordable cover can continue to be offered to customers.

A prime example of the danger of not responding to the signals provided by insurers occurs in flood cover, where a moral hazard can be created by inappropriate legislation. Where inappropriately low flood cover prices are mandated by government for property built on flood plains, those prices create a hazard. People in these properties are being put in harm's way, because inaccurate pricing of the risk leaves owners and residents unaware of the implications of building and living in such locations. Where the losses become too great or the costs of maintaining that cover outweigh returns from other lines, insurers may choose to abandon that market, leaving owners and residents without financial protection.

Legislation can be a driver for insurers to innovate and develop new products. The Carbon Reduction Commitment in the UK and the EU Environmental Liability Directive have led to a wealth of products being developed to help customers mitigate the potential risk exposures arising from the legislation. It is critical to understand the implications of new legislation that raises the importance of government dialogue with bodies such as the European Insurance and Reinsurance Federation and Association of British Insurers among others. Examples of legislative developments that need understanding by regulators and the industry include

- Extension of the Environmental Liability Directive to the marine environment. Land-based pollution is much easier to trace to source with more readily assessable remediation requirements—how this directive will be applied to a much more dynamic environment is not yet clear;
- Carbon sequestration is still an unknown technology with potential major risks. Mandated cover may create a moral risk again with an unknown liability; and
- Although there is currently no liability regime for carbon emissions (beyond the ETS), there may be potential for a precedent to be set. Liability covers for businesses, in particular emissions-heavy industries, may require additional capital to be set aside or the taking out of new policies to cover any lawsuits. The ability to establish a causal link between historic emissions of a single business or person to a specific weather-related event seems to be some way off at present.

In some developing countries lack of a legal framework can also make it impossible to conduct some forms of insurance. Where there is an absence of an effective judicial system it can make the processing of claims extremely difficult. It is vitally important that insurers can maintain their promises to customers in the event of a claim. For this reason a reasonable and balanced legal framework is important for insurers and in the best interests of all parties.

Technology as a Differentiator

The importance of technology to the insurance industry continues to grow. Most insurers are now able to assess risks using Geographic Information Systems (GIS) to calculate exposure to flooding, subsidence, and a whole host of risk factors. These tools allow customers to obtain an accurate assessment of the risks their property is exposed to and allows insurers to price accordingly. In the UK, most insurers can assess the risks down to the individual property level, allowing consideration of specific topographical variations or flood resilience measures within the property. As climate science improves, it can only help to inform insurers and customers on exposure to the direct impacts of climate change such as flooding or subsidence.

GIS is becoming vital in assessing risk accumulation, i.e., understanding how many customers or policies an insurer may have in a single location or area. Incidents such as earthquakes, flooding, or wind events can have wide-ranging geographical impacts. By mapping exposures geographically, the accumulation of risk can be managed so as not to jeopardize the balance sheets and to plan reinsurance cover effectively. However, at present there is a lack of models that can accurately predict localized impacts of climate change.

When the worst does happen, it is vital to respond quickly and effectively to customers in need. GIS allows emergency response procedures to kick in, facilitating understanding of where customers are, what policies they have, and what help needs to be put in place to assist them. These services can be a true differentiator among insurers as changing intensity and frequency of extreme weather events occur.

Specialists are developing skills in modelling weather-related impacts. Experts such as Risk Management Solutions (RMS) and the Benfield Hazard Centre provide models that can help insurers make pricing decisions. Brokers and reinsurers also work with insurers to provide modelling based on industry-wide claims data that benefit all insurers and customers. Having access to the best modelling can inform pricing to make sure risk is accurately assessed as well as making the right reinsurance decisions. For example, an insurer may decide to take out aggregate reinsurance cover to guard against a series of frequent smaller events like a succession of localized flood incidents. As reinsurance is bought on an annual basis, regular updating of these models is important and allows for improved climate science on an ongoing basis.

An area, more obvious to customers, that is providing real opportunities for insurers to differentiate themselves is provision of risk management information to customers. Nearly all insurers have some form of web-delivered risk prevention information, but mobile technology is providing new options. Some examples include

- Eurotempest provides email notifications of extreme wind speeds and gusts that could cause damage and can provide notifications up to five days in advance;
- Flood warnings can now be sent by mobile phone text message to customers to give them time to move property upstairs and take other flood-resilience measures; and
- In Santiago, RSA motor customers receive warnings to their mobile telephones on days when atmospheric pollution is high and certain cars cannot be used (RSA, 2009).

Consumer Behavior and Climate Change (Personal Lines)

As we have seen earlier, it is extremely unlikely consumers will desire to purchase multi-year policies guarding against general climate risk. A large part of consumer behavior is driven by cost and this situation is particularly the case in countries where aggregator websites operate.

Most consumer research around green products highlights that customers want first of all a cost-effective and quality insurance product covering traditional risks. If the product has green elements at no extra cost it could help improve retention or satisfaction. The perception of green product elements varies by country, sometimes driven by media coverage.

- In some countries green products are seen as having lower quality,
- Green products are often perceived as costing more and only a small segment of consumers are willing to pay extra. This behavior may be changing but is not clear with the financial crisis impacting consumers how much or how rapidly it might change,
- Internal Canadian research at RSA has indicated that customers do not value green elements if they are free but if the green elements are optional extras, they may be selected if there is a nominal extra charge.

The decision on buying personal lines of insurance can be described as a last-minute decision. When buying a home or a car, insurance can often be a last-minute consideration. This situation means the opportunity for an insurer to differentiate in this area is limited. Communication with the policyholder is possible, but caution needs to be exercised. Many customers view communications from the insurer as junk mail, perceiving it negatively or as competing against a range of other priority messages that need to reach the policyholder.

Motor

Over the past five years insurers have experimented with pay-as-you-drive models of insurance. Essentially all utilized GPS technology within the vehicle to track and assess behavior. Most offerings were based on carbon or safety goals but have since been withdrawn from the market (Insurance Daily and Montia, 2008). As GPS technology is not yet standard in cars, although increasing, the experiments required vehicles to have the appropriate technology. The costs of new installations and any maintenance costs usually resulted in the products not being cost-effective. In addition, the products were not usually targetted at mainstream consumer segments so lower returns were achieved. Once GPS technology is universal in vehicles, there will be a wider range of applications of pay-as-you-drive insurance, which will encourage and reward more carbon-friendly behavior.

Home

Although there is a general perception that cover for micro-renewables is an innovation needing to be added to homes policies, cover for micro-renewables is a part of the standard homes policy cover, as long as the micro-renweables are declared by the home owner. Micro-renewables generally increase risk to the property e.g., theft, fire risk, wind damage, etcetera. There has been innovation in the area of guaranteeing returns from micro-generation such as the tariff scheme provided by the local government in Ontario, Canada. One of the biggest opportunities for home insurance relates to the supply chain discussed below.

Supply Chain

The claims supply chain is where the role of the insurer becomes tangible. The materials used within a claim, the resilience of repairs to future weather impacts, and the carbon footprints of the chosen claims suppliers can provide a real opportunity for the insurer to differentiate itself and to contribute to mitigation and adaptation.

Many insurers are using reused, repaired, and recycled car parts for motor customers. While this practice is becoming increasingly common, there is still an issue around consumer perception that utilizing such components may compromise safety.

RSA is the first insurer globally to assess the carbon impacts of its home policy to the World Resources Institute/World Business Council for Sustainable Development draft carbon label protocol for products and services. Through a supplier collaboration agreement, a joint target to reduce carbon by 15 percent by the end of 2013 has been agreed (RSA, 2011).

There are still barriers to widespread adoption of sustainable materials within the claims supply chain as highlighted by ClimateWise (Meyricke and ClimateWise Sustainable Claims Steering Group, 2010). There is still a lack of understanding on what "sustainable materials" means and a perception that they increase costs. Many claims suppliers do not assess their provision of services on sustainable criteria nor do they assess their performance against such criteria, e.g., use of Forestry Stewardship Council timber, energy-efficient appliances, water-efficient devices, et cetera. This inattention is likely to decrease over time as more of the major insurers increase their activity in this area.

Interesting anecdotal evidence has also suggested that some customers affected by floods have not requested flood-resilient measures, e.g., electric sockets at waist height, so as not to impact the potential resale value of the property by signalling the danger of future floods to potential buyers.

A growing trend in claims is the frequency with which customers elect to take cash settlements. Cash settlements limit the insurers' ability to utilize a sustainable claims supply chain but incentives for customers to do so can be provided. Many insurers are providing cash incentives for the consumers to utilize energy-efficient appliances and other environmentally friendly goods.

Consumer Behavior and Climate Change (Commercial Lines)

Commercial insurance most commonly involves insuring businesses and property. Insurers' speciality business lines may focus on marine, renewables, construction, risk management, or small businesses, among many others.

One of the key differences between personal and commercial lines is the relationship of the insurer with the client. The relationship may be direct or through a broker, but the provision of commercial insurance is considered extremely early. For example, when constructing new retail premises or operating an office block, ensuring effective insurance cover is critical to enable construction to start or to have people in the building. The scale of loss is much bigger, so naturally a greater focus is placed on prevention and the insurance decision is given greater attention.

All these types of premises and activities need to consider the standard types of risks and business interruption as described in the risk section associated with climate change. Large businesses are also likely to be those exposed to legislation relating to climate change such as the UK Carbon Reduction Commitment. Insurers often have large risk management staffs helping commercial customers adapt to these legislative and best practice standards and these staffs can generate substantial consultancy income streams.

The claims supply chain plays a major role in commercial insurance. Many buildings are Building Research Establishment Environmental Assessment Method (BREEAM) or Leadership in Energy and Environmental Design (LEED) certified and so any claims require a like-for-like replacement to retain their rating, e.g., fitting the same energy-efficient wall insulation. Environmental issues also take on a new scale for large premises. For example, a burst pipe at the top of a 50-floor building can be more serious and damaging with a far greater waste of water. Environmental considerations are increasingly being demanded from insurers by customers who are leaders in the sustainability field. Ways of handling these considerations are becoming an increasingly common feature of commercial tenders for insurance.

Legislation has also driven the growth of renewable energy insurance because of carbon EU targets and national subsidies to stimulate climate-friendly energy generation. While such legislation is increasingly seen as a major climate-friendly opportunity for insurers it still accounts for a small portion of core business for globally diversified insurers.

Focusing on strategic issues

While not every issue relating to climate change can have a policy or customer-offering built around it, it is important for insurers to focus on the impact of climate change on core strategic issues and develop long-term strategies that will benefit their companies. Tackling long-term impacts can improve reputation, relations with responsible institutional investors, and a host of ratings indices, e.g., DJSI, FTSE4 good.

Insurers are collaborating more frequently on climate change issues in such groups as the insurance working group of UNEP-FI. By working together they get access to new ideas and knowledge to help mitigate shared risk. Such collaborations are also occurring throughout the insurance industry with brokers and reinsurers working together to improve understanding.

In addition to important cooperative endeavors there is also a growing competitive aspect to insurers' tackling of the strategic issues raised by climate change. The desire to be the leader in a particular field or to be able to demonstrate superior risk management knowledge is important. For example,

- Flooding is a common issue of concern, with insurers tackling it through a variety of different approaches such as awareness campaigns, being engaged in political events, creating flood-proof homes, and developing test sites for sustainable urban drainage schemes. The information researched and publicity developed help raise awareness about the importance of government spending on flood defences.
- The low-carbon economy is often cited as a requirement for a sustainable system. Many in the insurance industry address this need through skills development or researching specific areas such as low-carbon transport, renewables, or other priority sectors that would accelerate the transition.

By demonstrating a focus on researching long-term issues that can help develop sustainable business models, insurance companies can attract the interest of responsible investors. Forward-thinking institutional investors are engaging in climate change management, and while these organizations are currently few in number, insurance companies that are included in sustainability funds and whose shares are being bought by such leading investors are likely to become increasingly attractive to investors in general.

Investments

As mentioned earlier the EU Solvency II legislation has increased the focus on insurers holding a robust low-risk investment portfolio. It will be essential that insurers hold enough capital to cover the costs of claims. They must also invest their premiums effectively and prudently. A causal link between managing funds responsibly and an increased return on investment has not been shown unequivocally to exist—some studies show a positive link, others do not. Investment in clean technology funds is generally high risk due to the prototype technologies being used. Screening out certain types of sectors or companies could be considered but any reduction in the breadth of investments open to selection is generally believed to reduce potential returns of the fund manager, so specializing in particular sectors is discouraged by some experts.

Climate risk has started to be incorporated into some types of investments but still accounts for a relatively small part of the market. The majority of most investment portfolios will focus on A-rated assets.

Equity and property investments are considered to be more readily accessible to the exercise of responsible investment approaches. Ownership of shares allows voting on issues and a greater disclosure of information for the interested investor. Likewise the impacts on property can be assessed and managed to a certain extent.

When considering cash, fixed-income securities, and bonds, it becomes harder to determine how climate risk considerations might influence these portfolios. However, there has been some innovative research, usually confined to assessing risk in bonds in obvious sectors like energy.

Perhaps the best approach will be to have climate risk integrated into the overall rating of an asset. Since ratings agencies set recommended levels to assets based on their assessment of risk, it would seem logical that incorporating climate risk into these assessments would help all investors.

Conclusions

The overarching message of this chapter is that climate risks are very large, are having, and will continue to have, a significant impact on the insurance sector, both directly and indirectly. Insurers have various means of dealing with this risk, and some means are more socially acceptable and more effective than others. The insurance sector as a whole is not currently being as proactive as it needs to be and climate risk impacts a lot more of their business than insurers realize. However, there are also great opportunities to tackle this climate issue. If these risks and opportunities are tackled effectively a viable insurance sector will continue into the future and will contribute substantially to climate change mitigation and adaptation.

References

Africa Partnership Forum. 2007. *Climate change and Africa.* Berlin: Africa Partnership Forum.

Arnold, M. 2008. *The role of risk transfer and insurance in disaster risk reduction and climate change adaption* (Policy Brief). Stockholm: The Commission on Climate Change and Development.

Baum, F., and M. Fisher. 2010. Health equity and sustainability: Extending the work of the Commission on the social determinants of health critical public health. *World Health Organization Commission* 20(3): 311–322.

Bordoff, J. E., and P. J. Noel. 2008. *Pay-As-You-Drive auto insurance: A simple way to reduce driving-related harms and increase equity.* http://www.brookings.edu/papers/2008/07_payd_bordoffnoel.aspx (Accessed July 26, 2011).

Brown, C. 2007. Households may be left without flood insurance. *The Independent.* http://www.independent.co.uk/news/uk/politics/households-may-be-left-without-flood-insurance-396546.html (Accessed July 26, 2011).

Charpentier, A. 2008. Insurability of climate risks. [Proceedings Paper]. *Geneva Papers on Risk and Insurance-Issues and Practice* 33(1): 91–109.

Clements-Hunt, P., R. Fischer, J. Lopez, P. v. Flotow, L. Cleemann, A. Hummel, et al. 2011. *Advancing adaptation through climate information services: Results of a global survey on the information requirements of the financial sector.* Oestrich-Winkel & Geneva: Sustainable Business Institute (SBI) & UNEP Finance Initiative (UNEP FI).

Climate Change Policy & Practice. 2010. UNEP FI and Insurance Industry Release Statement on Climate Change Adaptation. http://climate-l.iisd.org/news/unep-fi-and-insurance-industry-release-statement-on-climate-change-adaptation/ (Accessed July 26, 2011).

ClimateWise. 2009. ClimateWise statement on the UNFCCC Copenhagen negotiations. http://www.climatewise.org.uk/storage/ClimateWise%20Copenhagen%20Statement.pdf (Accessed July 26, 2011).

Conning & Company. 1994. Lighting candles in the wind: Industry response to the catastrophe problem. Hartford, CT: Conning & Co.

De Loach, J. W. 2000. *Enterprise-wide risk management: Strategies for linking risk and opportunity.* London: Financial Times/Prentice Hall.

Earth Policy Institute. 2006. Plan B updates. Setting the record straight: More than 52,000 Europeans died from heat in summer 2003. http://www.earth-policy.org/plan_b_updates/2006/update56 (Accessed July 26, 2011).

Ernst & Young, and Oxford Analytica. 2009. *The 2009 Ernst & Young business risk report: The top 10 risks for global business.* London: Ernst & Young LLP.

Epstein, P. R., and D. Ferber. 2011. *Changing planet, changing health: How the climate crisis threatens our health and what we can do about it.* Los Angeles: University of California Press.

Few, R., M. Ahern, F. Matthies, and S. Kovats. 2004. *Floods, health and climate change:A strategic review* (Working Paper 63). Norwich: Tyndall Centre for Climate Change Research.

Gibbs, E., and J. DeLoach. 2006. Which comes first . . . Managing risk or strategy-setting? Both! *Financial Executive* 22(1): 34–39.

Hellmuth, M. E., D. E. Osgood, U. Hess, A. Moorhead, and H. Bhojwani. eds. 2009. *Index insurance and climate risk: Prospects for development and disaster management.* Climate and Society No. 2. International Research Institute for Climate and Society (IRI). New York: Columbia University.

Insurance Daily, and G. Montia. 2008. NU withdraws "pay as you drive" policy. http://www.insurancedaily.co.uk/2008/06/16/nu-withdraws-pay-as-you-drive-policy/ (Accessed July 26, 2011).

International Energy Agency (IEA) 2010. *Key world energy statistics 2010* Paris: OECD/IEA.

IPCC. 2007a. Summary for policymakers. In *Climate Change 2007: Impacts, Adaptation and Vulnerability. Contribution of Working Group II to the Fourth Assessment Report of the Intergovernmental Panel on Climate Change*, ed. M. L. Parry, O. F. Canziani, J. P. Palutikof, P. J. v. d. Linden, and C. E. Hanson.http://www.ipcc.ch/pdf/assessment-report/ar4/wg2/ar4-wg2-spm.pdf (Accessed July 26, 2011).

———. 2007b. Synthesis report. Contribution of working groups I, II, and III to the fourth assessment report of the Intergovernmental panel on climate change.

In *IPCC Fourth Assessment Report (AR4)*, ed. Core Writing Team, R. K. Pachauri, and A. Reisinger (p. 104). Geneva, Switzerland: IPCC.

Jóhannsdóttir, L. 2009. Woe awaits insurers! The Nordic insurance industry and climate change. In *Rannsóknir í félagsvísindum X*, ed. I. Hannibalsson. Reykjavík: Félagsvísindastofnun Háskóla Íslands.

Lessard, D. R., and R. Lucea. 2006. *Embracing risk as a core competence: The case of CEMEX*. Cambridge, MA: MIT Sloan School and George Washington University.

Leurig, S., and PricewaterhouseCoopers LLP. 2010. *The ripple effect: Water risk in the municipal bond market*. Boston: CERES.

Lloyd's 360 Risk Insight. 2009. *Climate change and security: Risks and opportunities for business*. London: Lloyd's.

Mechler, R. 2008. Climate insurance. http://www.iiasa.ac.at/Research/RAV/Projects/climate-ins.html (Accessed July 26, 2011).

Meyricke, R., and ClimateWise Sustainable Claims Steering Group. 2010. *Sustainable claims management*. London: ClimateWise.

Mills, E. 2003. Climate change, insurance and the buildings sector: Technological synergisms between adaptation and mitigation. *Building Research & Information* 31(3–4): 257–277.

———. 2005. Insurance in a climate of change. *Science* 309: 1040–1044.

———. 2006. Synergisms between climate change mitigation and adaptation: An insurance perspective. [(Invited). LBNL-55402]. *Mitigation and Adaptation Strategies for Global Change, Special Issue on Challenges in Integration Mitigation and Adaptation Responses to Climate Change* 12(5): 809–842.

———. 2007. *From risk to opportunity: 2007 insurer responses to climate change*. Boston: Ceres.

———. 2009. *From risk to opportunity2008: Insurer responses to climate change*. Boston: Ceres.

Mills, E., R. J. Roth, and E. Lecomte. 2005. *Availability and affordability of insurance under climate change: A growing challenge for the U.S.* Boston: Ceres.

Munich Re. 2011. *Natural catastrophes 2010: Analyses, assessments, positions*. Munich: Geo Risks Research (GEO/CCC1).

National Disaster Management Authority. 2010. Flood 2010 updates. http://ndma.gov.pk/flood-2010.html (Accessed July 26, 2011).

Perroy, L. 2005. Impacts of climate change on institutional investors assets and liabilities. Environmental Research Group of the Institute of Actuaries & Climate Change Capital. http://www.ica2006.com/Papiers/3108/3108.pdf (Accessed July 26, 2011).

Reiman, K. E., ed. 2007. *Climate change: Beyond whether*. Zurich: UBS AG, Wealth Management Research.

RSA 2009. *Corporate responsibility report 2009*. London: RSA Group.

RSA 2011. *Corporate responsibility report 2010*. London: RSA Group.

Stahel, W. R. 2010. Risk management, insurance and climate change. *Asia Insurance Review*, pp. 96–98. http://www.genevaassociation.org/pdf/News/1001%20AIR.pdf (Accessed July 26, 2011).

Swiss Re. 2010, September. Innovative insurance solutions are key to helping local communities finance the rising costs of climate risks. http://www.swissre.com/media/news_releases/Innovative_insurance_solutions_are_key_to_helping_local_communities_finance_the_rising_costs_of_climate_risks_says_new_Swiss_Re_publication.html (Accessed July 26, 2011).

The Carbon Trust. 2008. *Climate change—a business revolution? How tackling climate change could create or destroy company value.* London: The Carbon Trust.

The Geneva Association. 2009. *The insurance industry and climate change—contribution to the global debate.* Geneva: The Geneva Association.

Tomlin, B. 2006. On the value of mitigation and contingency strategies for managing supply chain disruption risks. *Management Science* 52(5): 639–658.

United Nations Environment Programme (UNEP). 2010. Insurers call for more action to adapt developing world to climate change. http://www.unep.org/Documents.Multilingual/Default.asp?DocumentID=647&ArticleID=6741&l=en&t=long (Accessed July 26, 2011).

U.S. Securities and Exchange Commission. 2010. SEC issues interpretive guidance on disclosure related to business or legal developments regarding climate change. http://www.sec.gov/news/press/2010/2010-15.htm (Accessed July 26, 2011).

CHAPTER 4

Competitive Advantages and Risk Management: Capitalizing on the Economic Benefits of Sustainability

Jonathan Smith and Efrain Quiros, III

Introduction

Climate change and ecological degradation are two of the many pressing global issues that make sustainability pertinent to all stakeholders, including shareholders. Such issues are relevant to companies across all industries and sectors and are increasingly pressuring managers to consider how their organizations can utilize resources effectively while minimizing impact on the environment. It is critical to consider the underlying relationship between business and the environment: companies ultimately operate within the natural environment and are dependent upon it for the resources they use. A narrow focus on short-term economic interests without consideration for environmental impacts ignores the overarching system in which the planet supports business; without resources from which to make products there are no products to sell and, consequently, no profits.

Business has a vested interest in preserving the very natural capital that its activities directly utilize and frequently, directly or indirectly, destroy. However, environmental concerns do not have to be the only focus for the sustainability agenda in business. In *Hot, Flat, and Crowded,* Thomas Friedman (2008) describes today's energy-climate challenge as a series of great opportunities disguised as seemingly insoluble problems. In this view, the pursuit of global sustainability provides managers with opportunities to improve business models and activities. Effectively understanding and applying approaches that contribute to a more sustainable world can enhance a wide range of trends, risks, and opportunities in the business environment.

This chapter reviews the fundamental roles and responsibilities of business and highlights the importance of the interdependent relationship between business and the environment. A review of environmental-business interactions and dynamics demonstrates that actively and creatively engaging the issues of global sustainability is necessary for managing business risks and for sensible business practice. The sustainability engagement experiences of two leading companies—HSBC and Walmart—provide examples of differing ways in which more sustainable business practices can be integrated into business activities to gain competitive advantages and economic benefits. As observed in these examples, approaches to sustainability engagement are far from generic and one-dimensional.

Business Responsibilities

In the macroeconomic view, business contributes to society via its position in the circular flow in the economy—supplying goods and services, utilizing factors of production (e.g., labor and capital), earning profits, and paying taxes (Mankiw, 2010). Considering these contributions, business indeed has a role in social welfare, even if a purely economic role. Economic responsibilities are fundamentally required of companies; however, the economic responsibilities are not the only ones. Companies have a range of responsibilities with economic, legal, and ethical implications—all of which have associated risks.

The fundamental role of business—to provide goods and services that consumers demand and to do so at a profit—drives *economic responsibilities*. The purpose of risk management is to mitigate unfavorable conditions that could expose the company to possible extra financial costs, and then to be proactive in reducing as many of these risks as possible in order to enhance the bottom line (i.e., profits).

Companies' obligations to comply with society's established rules as they participate in society's economy shape their *legal responsibilities*. Noncompliance with legal obligations may involve punitive monetary fines. However, reputational risks arising from the noncompliance can often have a greater impact as existing and potential future customers, suppliers, and perhaps even employees turn away from the company and its products and services. Companies need to assess the risks embedded in the legal environment and framework so they can plan and manage activities in a manner that will establish favorable positions in the competitive environment. Companies that adapt proactively to forthcoming regulations and changes in the legal environment can gain competitive advantage over competitors that must change reactively.

Ethical responsibilities are those expected of business by society beyond any regulatory, legal, or other compulsory framework. It is important that companies consider how various stakeholders perceive business activities, including customers and the public at large. Disregarding ethics can expose companies to brand risks and reduce consumption of their products and services—again effecting revenues and financial performance.

It is important that companies consider all forms of responsibilities so they can reduce risk exposure as much as possible and thus increase profits. Effective risk management can therefore align interests for the environment *and* for business. Business needs to be committed to sustainability and a sustainable world needs commitment from business. As Porritt (2007) points out, sustainability was once a desirability but is now a necessity. The key point to emphasize is that there are opportunities here for the reconciliation of interests.

The perspective of Milton Friedman (1970) that the only social responsibility of business is to utilize resources, within the ground rules set by societies and their governments, to increase profits strongly influences the views of many business leaders. Based on this view, critics of sustainability often argue that environmental business activities are inherently opposed to business interests. Drawing on Rayment and Smith (2010), we argue that the view that business and environmental interests are competing entities—in the sense that for one to gain the other must lose—is limited and misguided. On the contrary, we argue that Friedman's principles do not fundamentally oppose commitments and actions supporting sustainability because environmental issues actually create opportunities for businesses. In accordance with Milton Friedman's principles, it would be wrong to discount sustainability because sustainability can directly affect profits.

Corporations are accountable to various stakeholders that have the ability to influence company activities, including shareholders, management, and employees (Crane and Matten, 2010). Corporate managers in particular, as agents of shareholders, are legally responsible for making decisions and running corporations in the interests of owners (i.e., shareholders) (Bakan, 2005, p. 35). Corporate managers who utilize Milton Friedman's principles to reject a commitment to finding ways to contribute to a sustainable world have a flawed argument because evoking Milton Friedman's view in pursuit of profit maximization actually compels them to engage in sustainability-committed actions—as discussed, environmental issues create opportunities for and threats to financial gains. Consequently, corporate managers who reject sustainability despite its economic implications are not acting in the interests of shareholders; moreover, they actually act contrary to the very principles that they claim to advocate.

Commitments and actions that contribute to a sustainable world are more than an environmental necessity and a societal desire; sustainability-oriented actions are beyond corporate volunteerism and philanthropy and do not have to be limited to compassionate business practices. In a strict business sense, sustainability provides competitive advantages. Sustainability engagement should, and must, transcend all business activities and functions, including internal and external activities. Managers should be concerned with sustainability because, with short-term implications for profits and long-term implications for the very existence of business, environmental issues are in fact business issues. Profit-driven companies and managers concerned with maximizing value need to engage in sustainability-supportive actions because they can enhance company and shareholder value by creating opportunities to mitigate risks, reduce costs, increase market share, and gain competitive advantages.

Environmental Issues: Risks and Opportunities

Rockstrom et al. (2009) outline nine planetary boundaries that are causing unacceptable environmental change: climate change, ocean acidification, stratospheric ozone depletion, nitrogen and phosphorous cycles, freshwater use, changes in land use, biodiversity loss, atmospheric aerosol loading, and chemical pollution. Although the definitions of each boundary are beyond the scope of this paper, the implications are directly pertinent to business interests. The environmental issues create a number of pressures within the business environment including population growth, overconsumption, environmental degradation, biodiversity (as a resource) loss, and pollution—all of which threaten company activities and the overall productivity of the economy. Wall, Minocha, and Rees (2010) summarize the trends within the business environment as follows:

- *Environmentally conscious consumers*—increasing environmental awareness by consumers creates a market for "green" products. Even when core company products and services have limited environmental impacts, company relationships and activities across the value chain and supply chain can influence consumer demand.
- *Environmentally- and cost-conscious producers*—corporations can achieve cost reduction via more environmentally friendly operations, processes, and activities. Adopting sustainability focused strategies and processes can simultaneously meet higher environmental regulations and reduce costs.
- *Environmentally- and risk-conscious producers*—failure to manage environmental risk factors exposes corporations to adverse publicity, lost

revenue and profit, and further possible financial repercussions (e.g., reduced credit ratings).

- *Environmentally conscious governments*—governments are increasingly scrutinizing corporations, including environmental damage caused by business activities, which can lead to the imposition of fines or more regulation.

Environmental issues have implications for all business activities—including strategy, marketing, risk management, public relations, legal, finance, human resources, and leadership. Environmental risks are increasingly relevant and important to business and managerial decision-making; prudent and responsible risk-management planning should address environmental risks as it would any other business risk. Managers should actively manage their business in ways that contribute to global sustainability and seek to integrate environmental management strategies because environmental issues are in fact business risks. Lash and Wellington (2007, pp. 130–137) argue that environmental issues, particularly climate change, present the following risks to businesses:

- *Regulatory risk*—companies should act responsibly and immediately in assessing the possible effects of future legislation and establish favorable positions before rivals do.
- *Supply chain risk*—companies should evaluate vulnerabilities throughout their supply chain (i.e., upstream) and their distribution network (i.e., downstream).
- *Product and technology risk*—companies should recognize that some companies will fare better in exploiting new opportunities for environmentally friendly products and services and work to be among the more successful ones.
- *Litigation risk*—companies should recognize that companies with poor environmental records face the threat of lawsuits (similar to lawsuits that tobacco and pharmaceutical companies face) and make sure their environmental records do not expose them to similar risks.
- *Reputational risk*—companies should be aware that they face judgment in the "court of public opinion." However, they should take advantage of the fact that, as with other risks, reputational risks can be opportunities for altering public opinion in positive ways.
- *Physical risk*—because the physical (natural) environment directly affects the overall operating environment for businesses, companies need to take note that physical conditions (e.g., droughts, floods, and storms) can adversely affect the entire value chain and develop appropriate protective strategies and actions.

These pressures and risks are creating a paradigm shift in the way businesses perceive environmental issues. Esty and Winston (2009) assert that environmentally friendly strategies provide businesses with added degrees of freedom to operate, profit, and grow. Bisson, Stephenson, and Viguerie (2010) contend that "*systematically spotting and acting on emerging [trends] helps companies to capture market opportunities, test risks, and spur innovation.*" Although managers do not single-handedly shape forces and trends in the business environment, they disregard the dynamics of climate change and the need for global sustainability at their own and their companies' peril. It is shortsighted to view resource consumption and environmental pollution as economic externalities unrelated to business activities, or as marginal costs to be "paid for" by society and the environment. As climate change continues to cause environmental problems with adverse implications to people around the world, national governments and intergovernmental organizations will actively pursue methods of quantifying environmental costs and seek to regulate activities that have adverse environmental affects.

In contrast to the old paradigm that pitted environmental interests and business interests against one another, environmental and business interests in the new global business paradigm are now aligned and interdependent. Management of environmental impacts is vital to companies due to increased levels of accountability and the need to minimize risk exposure from business activities that adversely affect the environment. The implications and applications of both sustainability and risk management transcend all business activities, internally and externally, from high-level management strategic planning to the execution of operating functions. The paradigm shift that brings environmental issues to the forefront of business attention strengthens the association between sustainability and risk management. Companies gain competitive advantages from taking responsible, proactive approaches to risk management. Perhaps no one has phrased it quite as well as Molak (1997, p. 3): "only in an unenlightened society are environmental safeguards mistakenly considered as being opposed to business interests and free markets." Risk management and sustainability relate conceptually and in application; both are economically sensible practices that can contribute to improving business models. In effect, risk management and sustainability are one in the same—mitigating environmental risks, as with any business risk, is a prudent and instrumental business activity.

Environmental concerns remain a priority for activists, conservationists, and scientists alike and they are increasingly a priority for managers. While scientists debate the extent and causes of climate change and policy makers debate what to do about climate change, the effects of climate change on businesses and the need for sustainability-focused actions by businesses are tangible and significant. As Packard and Reinhardt (2000) contend, the impact of global warming on business is real and there are no excuses for inaction. This paradigm shift has made including sustainability factors imperative in managerial decision-making (Stoner and Wankel, 2010).

Managing for Global Sustainability

The United Nations' Brundtland Commission (World Commission on Environment and Development, 1987) presents sustainable development as having three pillars: economy, society, and environment. This view can limit the concepts and applications of sustainability if it is interpreted as depicting each entity as independent. Alternatively, Giddings, Hopwood, and O'Brien (2002) emphasize that the relationship between business and the environment is *dependent*: business is an entity within the economy, the economy exists within society (as a societal construct), and society exists within the natural environment. This holistic view more adequately describes the interconnected relationship between business and the environment, establishing that companies are ultimately dependent on the natural environment.

Environmental concerns, including climate change, are fundamentally important to companies because economic interests are inherently dependent on the environment for factors of production—e.g., labor and natural resources. In this view, companies have the ability, and are obligated, to supply products sustainably *and* to create demand for sustainable products (Hart, 1997). The authors believe environmental stability and sustainability engagement are in the long-term interests of business, and agree with Hart's (1997, p. 121) assertion that "the responsibility for ensuring a sustainable world falls largely on the shoulders of the world's enterprises." Managers would be misguided to believe that the environment and business have independent interests, and that effective strategic business planning can disregard environmental concerns. A focus on short-term economic motives without regard to long-term environmental impacts fails to recognize the fundamental dependency of business on the environment.

The concept of the triple-bottom-line (TBL), commonly referred to as "*people, planet, and profits*," proposes that sustainability provides opportunities for win-win-win situations—simultaneous economic, social, and environmental benefits. Savitz and Weber (2006) define a sustainable company as "one that creates profit for its shareholders while protecting the environment and improving the lives of those with whom it interacts." Furthermore, the Dow Jones Sustainability Indexes (2009) define corporate sustainability as "a business approach that creates long-term shareholder value by embracing opportunities and managing risks deriving from economic, environmental and social developments." Sustainability is fundamentally about reconciling interests, and the benefits of sustainability are beyond an all-or-nothing perspective. Economic interests cannot be ignored because without financial viability, a business operating in a market economy will inevitably cease to exist. As Fox (2007) claims, the primary concern of managers is to be economically responsible because companies cannot pursue environmental and social initiatives without first being economically sustainable.

In the long term, companies that respond most appropriately to changes in the business environment survive and thrive, those that do not perish. Sensible strategic planning is required for companies to be in the former group. Although not every company can be considered a "green company," managers should strive to make their companies as "green" as possible. At the very least, companies are compelled to integrate sustainability into their strategies and operations as much as their business functions and activities will allow. Reinhardt (1999) proposes that companies can actively utilize environmental strategies to *differentiate* themselves by offering greater environmental benefits than those of competitors, *manage the competitive environment* such that regulations and standards adversely affect competitors in greater proportion, *reduce overhead expenses* by cutting costs and improving environmental performance, and *mitigate risks* by reducing exposure before adverse events can create financial liabilities.

To enhance shareholder value, companies can capitalize on market demand for sustainability-associated products and services. Lovins, Lovins, and Hawken (1999) propose four specific shifts in business practices that companies can utilize to integrate "natural capitalism":

1. *Increase the productivity of natural resources*—reduce waste in current production processes and generally utilize resources more efficiently. Doing so provides the opportunity to reduce costs and therefore to increase profits.
2. *Develop environmentally friendly production models*—implement manufacturing processes that prevent waste or utilize waste for other purposes (e.g., create new products, processes, and revenue sources). Using such production models provides the opportunity to reduce costs and/or generate revenues, and therefore to increase profits.
3. *Adopt a solutions-oriented business model*—rather than simply selling goods, refocus business purposes and objectives on providing solutions (e.g., providing lighting instead of selling light bulbs). This new focus aligns producer and consumer interests, and creates mutually beneficial

opportunities by simultaneously rewarding resource efficiency and productivity. The greater efficiency and productivity reduces costs and generates revenues, and therefore increases profits.

4. *Reinvest in natural capital*—restore, sustain, and expand the planet's ecosystem because natural resources are essential to the supply of goods and services. Without products there are no revenues, no profits, and businesses are nonexistent.

Environmental concerns are indisputable factors within the business environment. The issues, challenges, and developments directly affect competitive advantage and should be addressed at the highest levels of corporate planning and management—particularly with regard to corporate strategies and risk management. Ultimately, corporate planning is "an attempt to decide how best to respond to, or anticipate, change" (Argenti, 1980, p. 10). In this case, the changes are in the form of environmental issues within the business environment, affecting supply and demand. Managers, as decision-makers, have the capacity to influence whether dynamics in the business environment will be opportunities to increase efficiency and reduce environmental risk exposure, or will become competitive advantages lost to rivals.

With economic responsibilities to shareholders, managers are compelled to consider sustainability engagement as a risk-management tool with implications for overall company value. Sustainability is now firmly on the corporate agenda—and is ever more important to business and management because it is the "right thing to do" *and* because it pays off financially (Brown and Turner, 2008). According to Elkington (1994), companies have little choice but to get involved in sustainability to be successful. From Elkington's initial proposition of "win-win-win" strategies, academics and practitioners alike have developed theories and applications for answering questions about why and how companies can utilize sustainability-focused approaches effectively to gain competitive advantages (i.e., eco-advantages) and, in doing so, enhance the triple-bottom-line. Epstein (2008) contends there is a strong and positive link between sustainability and corporate value—noting that companies utilize sustainability initiatives to decrease operating costs and increase revenues, thereby gaining the following benefits:

- *Financial payoffs*—e.g., reduced operating costs (including lower litigation costs), increased revenues, lower administrative costs, lower capital costs, and stock market premiums;
- *Customer-related payoffs*—e.g., increased customer satisfaction, product innovation, market share increases, improved reputation, and new market opportunities;

- *Operational payoffs*—e.g., process innovation, productivity gains, reduced cycle times, improved resource yields, and waste minimization; and
- *Organizational payoffs*—e.g., employee satisfaction, improved stakeholder relationships, reduced regulatory risk, and increased learning.

Lash and Wellington (2007) suggest companies can improve "climate competitiveness" with the following basic steps: measuring the environmental impacts of business activities, identifying risks and opportunities, adapting business responses, and doing the former steps *better than competitors*. These competitive endeavors are a real dynamic in the business environment and managers can either utilize these activities to gain competitive advantages or risk losing competitive advantages to rivals that do. Opportunities are never completely lost; someone capitalizes on them. In business, it is either your company or a competitor.

Sustainability Engagement Approaches

In *Green to Gold*, Esty and Winston (2009) assert that all business activities fit into eight green business strategies, which accomplish four strategic tasks, which ultimately fall under two general categories: "managing the downside" or "building the upside." See Table 4.1. Utilizing the framework of Esty and Winston as an outline, one of the authors developed a study to research the sustainability engagement of two industry-leading companies that are widely recognized for sustainability efforts: Walmart and HSBC (Quiros, 2011). Looking beyond why companies commit to working toward a more sustainable world, the research focused on what companies were doing and how they were benefitting: i.e., what successful sustainability engagement effectively entails. The study evaluated sustainability communications and categorized sustainability activities under the following schematic with Objectives I-II, Strategies A-D, and Tactics 1–8:

The research highlights that Walmart and HSBC, while both recognized as sustainability leaders, have approaches to sustainability that vary in objective and execution—both implicitly or explicitly. Internalized strategies can be implemented quickly and provide tangible and measurable benefits (e.g., increased revenues and direct cost-savings). Alternatively, it can take longer to realize the benefits of indirect, external strategies (e.g., ones focused on company reputational value and risk management), with benefits that are less tangible and more difficult to quantify. Walmart provides an example of active engagement of sustainability, with an internal focus aimed at controlling costs—integrating sustainability into activities across its value chain. In

Table 4.1 Sustainability objectives, strategies, and tactics

Objective	*Strategy*	*Tactic*
I. **Controls** "Managing the Downside"	A. **Costs** Reducing operational costs and environmental expenses	1. **Efficiency**—improving resource productivity and usage by enhancing efficiency—e.g., waste elimination or utilization, pollution prevention, and energy conservation. 2. **Regulations**—decreasing disposal costs, regulatory fines, compliance expenses, etc.—e.g., eliminating costs associated with waste disposal and pollution control, including time and money spent. 3. **Value-chain engagement**—capturing the value of reduced environmental burdens throughout primary, secondary, internal, and external functions; to include supply-chain management (when not referring to risk management specifically)—e.g., more efficient packaging, supply-chain efficiency, maximizing usage of space, sourcing, and distribution.
	B. **Risks** Identifying and reducing risk exposure	4. **Risk management**—anticipating and addressing issues, regulations, and mandates; ensuring compliance and minimizing exposure to financial, strategic, operational, and hazard risks—e.g., liability, competitive, consumer, supply-chain, regulatory, natural, and employment risks.
II. **Opportunities** "Building the Upside"	C. **Revenue** Driving revenues by providing products that meet demand	5. **Product design**—making environmentally friendly products—i.e., designing, redesigning, modifying, or improving current products to meet the demands of green-conscientious consumers. 6. **Sales and marketing**—building competitive positions for current products or generating consumer loyalty based on sustainability and green attributes. 7. **Growth**—driving growth by promoting innovation—i.e., developing new products or new markets that are completely different from current product purposes or geographic markets.
	D. **Intangible** Creating value with overall corporate sustainability	8. **Value**—building corporate reputation and brand value—e.g., share-price performance, customer loyalty, brand image, access to capital.

Source: Adapted from *Green to Gold* (Esty and Winston, 2009).

contrast, HSBC provides an example of a passive engagement of sustainability, with an external focus aimed at creating opportunities—allowing the company to generate revenues and promote a benevolent brand image with limited engagement of internal business operations.

The research found Walmart's sustainability engagement primarily focuses on implementing controls and, in particular, managing costs (Quiros, 2011). A consumer goods retailer could be tempted to focus sustainability efforts on exploiting opportunities to maximize revenues instead of implementing cost-reduction practices. Walmart's active integration of sustainability is beyond passive compliance with expectations and standards; without legal or regulatory obligations to do so, Walmart actively changed its business model and operations to reduce environmental impacts and subsequently gained economic benefits. This approach seeks to reduce costs by increasing energy efficiency in its retail outlets and enhancing efficiencies throughout its value chain. According to Quiros' (2011) research, Walmart's internalized approach to sustainability engagement prioritizes enhancement of its business model more than enhancement of its brand image. Walmart's brand recognition for sustainability is a result of internal implementation, not external marketing.

Walmart focuses on cost controls but also pursues revenue growth and value creation, reflecting a comprehensive understanding of how sustainability can be effectively utilized to capitalize on opportunities by both reducing costs and increasing revenues. With a vast number of large retail outlets, Walmart has significant overhead costs. As such, it is sensible and expected that Walmart seek reductions in variable overhead costs whenever and however possible. This approach is very utilitarian in which sustainability is actively integrated into internal activities and operations. It is always important for a company to increase revenues and capitalize on opportunities; however, with an extensive value chain, supply chain, and distribution network, Walmart's comprehensive integration of sustainability initiatives throughout its business activities is very economically prudent. Furthermore, these initiatives are representative of an approach that is more proactive than risk-averse.

The second organization studied in Quiros' research is HSBC. The research found HSBC's sustainability engagement to be generally broad and abstract—primarily utilizing tactics associated with risk management and value creation that are generally intangible but nonetheless important. HSBC's sustainability approach is largely external, focusing on activities that do not directly affect internal operational functions such as cost controls, energy efficiency, or resource productivity. This emphasis does not suggest the external approach is insincere. HSBC neither utilizes natural resources for manufacturing nor sells physical products. As such, HSBC does not need to adapt products or improve processes to be more "green" or "sustainable" as companies do in industries with greater direct environmental impacts.

HSBC implements its sustainability initiatives by managing risks in its investments and driving revenues from those investments. HSBC establishes sustainability standards and expectations for partners when making investment decisions. HSBC exerts its influence to drive sustainability principles throughout its "supply chain" to gain two distinct, albeit indirect, benefits: (1) improving the sustainability practices of partners subsequently reduces risk exposure to HSBC investments and (2) HSBC gains positive brand association and value without actually integrating sustainability into its internal operations. HSBC also utilizes its insights into sustainability-focused techniques to create new investment opportunities—driving revenue growth with new types of investments associated with sustainability. This indirect engagement allows HSBC to reduce risk exposure and simultaneously project a positive, CSR-friendly brand image to all stakeholders by highlighting its sustainability engagement.

HSBC's sustainability engagement, although external and risk-averse, appeals to sustainability-conscious consumers and simultaneously drives revenues via strategic investments. Although the approach is not necessarily comprehensive, being generally external for opportunities and controls, HSBC's ability to capitalize on its sustainability engagement suggests a comprehensive understanding of the subject and applications. In addition, HSBC continues to receive recognition for sustainability efforts with its strategic external approach. As a financial services company, HSBC's business model maximizes revenues through investments. Thus, it is prudent for HSBC to utilize an approach that focuses on making investments in sustainability-related industries and companies. CSR and sustainable-investing are growing market trends, and a financial investment company should utilize all trends to make profitable investments. However, HSBC could also incorporate sustainability into internal operations to improve efficiencies and gain additional competitive advantages associated with lower operating costs. Internal engagement (i.e., controls) can be utilized to compliment external engagement (i.e., opportunities).

Walmart sells green products and has a positive brand image associated with sustainability, yet its approach remains cost-oriented. This approach is consistent with Walmart's business model of reducing costs as much as possible so it can offer lower prices to customers, preferring cost-leadership to differentiation. Walmart's approach is tangible and provides the ability to realize gains in the short term, whereas HSBC's approach is more abstract and intangible. Walmart's needs are intrinsically different from those of HSBC because the companies operate in different industries and offer different products. Consequently, their respective sustainability approaches are, and should be, different.

A normative view of sustainability engagement recognizes the importance of environmental issues but does not necessarily incorporate sustainability into internal functions. Companies that adopt a normative approach to sustainability may be keen to project an image reflecting the importance of sustainability—possibly because they are risk-averse—but may not understand why or how to incorporate sustainability into operations. The utilitarian applications of sustainability are increasingly beneficial, dependent on conditions, factors, and context in which a company is operating. It is important to have a normative understanding of sustainability, but also to adopt a utilitarian approach to maximize the benefits of engaging actively with the need for a more sustainable world.

Companies' operating environments, activities, and purposes vary from business to business; thus, conditions for decision-making vary. Likewise, approaches to sustainability vary, depending on the respective context and conditions in which companies individually operate. Companies and business models cannot apply any single, generic sustainability approach; nor should they. Limiting engagement to a single tactic or strategy limits the potential benefits of sustainability engagement. Managers need not seek an approach that is the same as that of another company; approaches will be inherently different depending on the context of business-operating environments. For managers, sustainability engagement provides the ability to maximize economic and environmental benefits by incorporating strategies to capitalize on both opportunities *and* controls. Sustainability commitment has a genuine economic rationale with its ability to increase revenues and/or reduce costs.

Walmart and HSBC provide examples of effective and successful approaches to sustainability engagement, with various combinations of internal and external strategies. Ultimately, the win-win of economic and environmental benefits is the most persuasive business rationale for sustainability. The effectiveness of sustainability engagement depends on how well it achieves the purposes of reducing environmental impacts, while also enhancing the bottom line. Rather than a generic, one-size-fits-all model, sustainability approaches and engagement can, and should be, conditional, situational, and circumstantial. Therefore, managerial decision-making about sustainability engagement should also reflect variability in respective conditions.

Strategic Sustainability Engagement

Environmental issues are part of the business environment, and it has become increasingly clear that companies must address changes in the business environment by developing strategic approaches to modify and develop activities throughout their value chains to capitalize on the various benefits of sustainability, or lose competitive advantages to rivals that do. In *Strategies for Diversification*, Ansoff (1957) outlines four opportunities for growth: market penetration, market development, product development, and diversification. These strategies provide a framework in which to summarize the opportunities that sustainability can provide; successful companies can utilize a combination of objectives, strategies, and tactics to capitalize on the benefits of sustainability.

Market penetration entails increasing sales without changing the product or market, by increasing sales volume to existing customers and/or targeting competitors' customers. In current markets, sustainability provides the ability to improve customer loyalty and attract more customers to current products. In pursuit of market share, companies can benefit from superior brand image and better corporate value. This approach is an effective method of utilizing product differentiation based on environmental-oriented attributes to attract customers, rather than directly attacking competitors. Moreover, the opportunities to reduce risks, control costs, and enhance the bottom-line that sustainability provides enhances the ability to gain cost-advantages over rivals. Companies can in turn leverage these competitive advantages to develop price-competitiveness or differentiation, depending upon overall corporate strategies and brand positioning.

Market development entails seeking new customers (in different demographics or geographic locations) for existing products or seeking new purposes for existing products. This strategy allows for access to a high volume of new customers with minor product modifications. With varying degrees of environmental concerns and policies among countries, companies can utilize sustainability engagement, product knowledge, and green strategies to develop new markets. The combination of low environmental impact and a positive brand image can help companies gain access to new geographic markets where new regulations may make domestic companies less competitive.

Product development entails offering products with different characteristics from the current product line, while competing in the same markets with the same product purposes. These product changes utilize excess production capacity, counter competition, adopt new technology, or protect market share (Lynch, 2006). Developing new products that are more environmentally friendly (i.e., differentiated), combined with lower overhead costs and prices, can create opportunities for differentiation or cost-leadership; thereby allowing companies to increase or protect market share and gain competitive advantages.

Diversification entails offering new products in new markets, in a complete deviation from present products and markets. Sustainability in itself need not be the fundamental purpose or raison d'être of business. Rather, sustainability compliments and enhances business activities. This observation is in line with the claim by Andersen, Ansoff, Norton, and Westen (1959) that "diversification is no substitute for exploiting opportunities in present product areas."

As we found in our research, sustainability approaches are, and should be, conditional depending on the dynamics in companies' respective business environments (Quiros, 2011). Sustainability engagement is far from generic and its range of applications allows for tailored approaches to specific business objectives or functional activities according to circumstances. To realize the full benefits of sustainability engagement, a comprehensive understanding and approach should consider the various stakeholders involved and be incorporated throughout companies' value chains. By utilizing various combinations of objectives, strategies, and tactics, successful companies can reduce environmental business risks and gain competitive advantages, increase revenues and/or reduce costs, and enhance overall company value—to the benefit of all stakeholders, financial and nonfinancial alike.

Conclusions

Business has economic roles and responsibilities that are integral to economic development. The new paradigm in the global business environment makes sustainability imperative for business. Whereas the old view pitted environmental interests against business interests, the new paradigm links these two "interest groups" as interdependent with mutual interests. In addition to the economic benefits, companies should seek to incorporate sustainability, wherever possible, for the sake of the natural environment upon which they are dependent. Short-term profits at the expense the natural environment jeopardize the long-term existence of companies.

Managers, with responsibilities to shareholders, are compelled to consider economic interests when making decisions. However, this commitment does not prohibit managers from considering environmental issues such as climate change and ecological degradation. It is because of obligations to shareholders that managers must consider *all* dynamics in the business environment with underlying economic implications—including environmental issues. Companies cannot discount the risks of climate change and forgo the opportunities to undertake sustainability-focused initiatives because taking a proactive approach can protect company viability and provide competitive advantages. Managers should not reject the concept of sustainability or "green business" based on misinformed beliefs that sustainability is inherently opposed to business interests.

Sustainability engagement should include analysis and managerial decision-making like any other form of corporate planning: to create competitive advantages that ultimately deliver positive returns, by capitalizing on opportunities and/or reducing risks. An approach to the management of environmental issues should be similar to the management of any other dynamics in the business environment that pose threats and risks. Due to the current and future risks that environmental issues create in the business environment, it is important that companies actively integrate sustainability-focused initiatives into business activities as much as possible. Companies with business models that do not allow for reduction of environmental impacts will ultimately face an operating environment in which they cannot compete. Sustainability and green business strategies provide methods for companies and managers to fulfill economic responsibilities while minimizing environmental impacts.

Sustainability generates value for business and the environment, and the convergence of environmental interests and business interests is ultimately important to all stakeholders. Sustainability-focused thinking should not be an alternative to standard business thinking, but instead it should be incorporated within standard business thinking. Although business interests are not fundamentally against environmental interests, too many business leaders do not understand or appreciate the value of the environment to their interests. Business leaders need better quality information about the link between business and the environment, and furthermore about the interdependent relationship between the two entities. Sustainability leadership is multidimensional and must engage various perspectives, interests, and responsibilities. Without comprehensive understanding, managers may not want to commit to and may be able to realize the economic benefits of sustainability engagement, to the detriment of the shareholders and all other stakeholders alike. Sustainability is no longer limited to a view in which it is opposed to business interests; it has become and is increasingly essential to business interests.

References

Andersen, T. A., H. I. Ansoff, F. Norton, and J. F. Westen. 1959. Planning for Diversification Through Merger. *California Management Review* 1(4): 24–35.

Ansoff, H. I. 1957. Strategies for diversification. *Harvard Business Review* 35(5): 113–124.

Argenti, J. 1980. *Practical corporate planning.* London: George Allen & Unwin Ltd.

Bakan, J. 2005. *The corporation: The pathelogical pursuit of profit and power.* London: Constable & Robinson Ltd.

Bisson, P., E. Stephenson, and S. P. Viguerie. 2010, June. Global forces: An introduction. *McKinsey Quarterly.*

Brown, M., and P. Turner. 2008. *The admirable company.* London: Profile Books Ltd.

Crane, A., and D. Matten. 2010. *Business ethics.* Oxford, UK: Oxford University Press.

Dow Jones Sustainability Indexes (DJSI). 2009. *Corporate sustainability.* http://www.sustainability-index.com/07_htmle/sustainability/corpsustainability.html (Accessed July 26, 2011).

Elkington, J. 1994. Towards the sustainable corporation: Win-Win-Win business strategies for sustainable development. *California Management Review* 36(2): 90–100.

Epstein, M. J. 2008. *Making sustainability work: Best practices in managing and measuring corporate social, environmental, and economic impacts.* Sheffield, UK: Greenleaf Publishing Limited.

Esty, D. C., and A. S. Winston. 2009. *Green to gold: How smart companies use environmental strategy to innovate, create value, and build competitive advantage.* Hoboken, NJ: Wiley.

Fox, A. 2007. Corporate social responsibility pays off. *HR Magazine* 52(8): 42–47.

Friedman, M. 1970, September 13. The social responsibility of business is to increase its profits. *The New York Times Magazine.*

Friedman, T. 2008. *Hot, flat, and crowded.* New York: Picador.

Giddings, B., B. Hopwood, and G. O'Brien. 2002. Environment, economy and society: Fitting them together into sustainable development. *Sustainable Development* 10: 187–196.

Hart, S. L. 1997. Beyond greening: Strategies for a sustainable world. *Harvard Business Review* 75(1): 66–77.

Lash, J., and F. Wellington. 2007. Competitive advantage on a warming planet. *Harvard Business Review* 85(3): 94–102.

Lovins, A. B., L. H. Lovins, and P. Hawken. 1999. A road map for natural capitalism. *Harvard Business Review* 77(3): 145–158.

Lynch, R. 2006. *Corporate strategy* (4th ed.). Essex, UK: Pearson Education Limited.

Mankiw, N. G. 2010. *Macroeconomics.* New York: Worth Publishers.

Molak, V. 1997. *Fundamentals of risk analysis and risk management.* Boca Raton, FL: CRC Press LLC.

Packard, K. O., and F. L. Reinhardt. 2000. What every executive needs to know about global warming. *Harvard Business Review* 78(4): 129–135.

Porritt, J. 2007. *Capitalism: As if the world matters.* London: Earthscan Publications Limited.

Quiros, E. 2011, February. Corporate sustainability: Profit-maximisation and competitive advantage. *Dissertation, Master of Science International Business Economics.* Cambridge, UK: Ashcroft International Business School, Anglia Ruskin University.

Rayment, J., and J. Smith. 2010. *MisLeadership: Prevalence, causes and consequences.* Farnham, UK: Gower Publishing Limited.

Reinhardt, F. L. 1999. Bringing the environment down to earth. *Harvard Business Review* 77(4): 149–157.

Rockstrom, J., W. Steffen, K. Noone, A. Persson, F. S. Chapin III, E. F. Lambin, et al. 2009. A safe operating space for humanity. *Nature* 461(24): 472–475.

Savitz, A. W., and K. Weber. 2006. *The triple bottom line: How today's companies are achieving economic, social, and environmental success—and how you can too.* San Francisco, CA: Jossey-Bass.

Stoner, J. A. F., and C. Wankel. 2010. *Global sustainability as a business imperative.* New York: Palgrave Macmillan.

Wall, S., S. Minocha, and B. Rees. 2010. *International business.* Harlow, UK: Pearson Education Limited.

World Commission on Environment and Development. 1987. *Our common future.* Oxford, UK: Oxford University Press.

CHAPTER 5

Climate Change Impact on Procurement: Risks and Opportunities

Lydia Bals

Introduction

Weather anomalies are occurring around us today. Although some people still doubt whether we can truly speak of an accumulation of extreme events more than in previous times and whether this is truly induced by human-caused climate change, some facts are astonishing. For example, the summer of 2010 was the hottest in Europe for 500 years and according to multi-model experiments, the probability of experiencing a summer with "mega-heat waves" will increase by a factor of five to ten within the next 40 years (Barriopedro, Fischer, Luterbacher, et al., 2011). Moreover, scientists estimate that in California a hypothetical superstorm could produce up to 10 feet of rainfall over 40 days, wind speeds of 125 mph, and cause damages between $300 billion and $1 trillion (Campbell, 2011). Such climate-disruption events can cause extreme effects on our modern infrastructures.

Although this chapter focuses on ways climate change may manifest itself as extreme weather conditions and events such as floods, dry periods, and storms that create supply chain disruptions, most of the considerations discussed will hold true also for environmentally or politically induced disasters in general. Nevertheless, this focus on climate is important for establishing the link to the sustainability issue, which lies at the heart of the solution-oriented focus of this chapter.

To shed light on the issues and approaches to solutions in this context from the perspective of procurement, this chapter addresses two basic questions: (1) What implications does climate change have for the procurement function and for supply chains? (2) What implications will climate disruptions have for what it means to be a "sustainable business" and for managing the procurement process in such businesses?

With increasing impacts from climate change, supply markets will become more heavily affected. However, it is important to distinguish between the goods market and the services market because the implications will be different for the two types of markets. Whereas goods, as well as physical inputs in general, are subject to logistical/physical flow considerations and natural resource market considerations, services are not. Therefore, the two will be discussed separately. The first question will be tackled in the next section to lay out the different risks implied in the two types of markets. Potential mitigation strategies will then be discussed.

The second question, which deals with the meaning of global sustainability in a business, brings out the greater context in which procurement needs to operate: What does "sustainable business" imply for procurement and how can procurement contribute as a business function? In addressing this question, the chapter will describe the facets of sustainability and their relation to procurement. The simplest interpretation of procurement's role in a sustainable business involves ensuring supply security so the company is able to produce its products and/or services. This role as a risk-reducer is a critical one, but it does not suggest the full potential of procurement's possible contributions to a sustainable enterprise. The discussion then moves on to consideration of procurement's full potential, not only as a risk-reducer, but also as an opportunity-spotter, especially by creating cross-functional collaboration. Finally, the chapter will summarize the discussion, the conclusions reached, and the outlook for the role of procurement in companies committed to aligning their operations with the need for global sustainability.

Climate Change Implications

Production-Oriented Businesses

For production-oriented businesses, climate change is already impacting natural resource markets, leading to higher market prices and greater price volatilities. Supply distortions caused by natural disasters such as floods (e.g., flooded coal mines in Australia and losses of wheat production in many countries) and subsequent effects on the supply chain (e.g., resulting increases in steel prices and food prices) are forcing companies to cope with insecure supply situations—supply situations that could potentially have severe implications for their own production. Such situations are especially present in industries with (close to) zero stock policies, such as the automotive industry. For food, the link to climate change has been shown clearly by a recent study (Lobell, Schlenker, and Costa-Roberts, 2011) that included calculations of some of the impacts of climate change on the worldwide production of corn and wheat yields since 1980. The study suggests that without global warming, total harvests of both crops would have been significantly larger. The estimated annual shortfalls amount to an approximately 23 million metric tons of corn in Mexico and approximately 33 metric tons in France. Relatively speaking, these figures amount to a decline in production of maize and wheat by 3.8 percent and 5.5 percent, respectively. One of the countries with the largest crop loss was Russia, where wheat production fell some 15 percent. The study estimates that the global drop-off in production may have caused a 6 percent hike in consumer food prices since 1980, some $60 billion per year (Lobell, Schlenker, and Costa-Roberts, 2011).

Even if resources are still available, the increased volatility of supply attracts speculative investment that aggravates price fluctuations apart from those caused by normal changes in supply and demand.

These effects are even more pronounced in industries dealing with more perishable inputs, such as the food industry, than in industries with durable inputs. Time is a particularly critical parameter where perishable items are involved. Distortions to physical transportation coming from natural disasters such as floods, where road transportation could be disrupted, can have devastating effects on companies. Hurricanes and blizzards could interrupt air and sea transportation, which can have similar impacts on companies. The impacts of those types of situations are usually more severe than in industries using durable inputs, where supply disruptions may involve a "mere delay" when one intermediate step in the chain fails. The rising food prices that are currently observable highlight the seriousness of impacts on perishable products. Food prices are on a "meteoric climb" (Reddy, 2011) and higher food prices are currently forecasted as being likely to continue (Schneider, 2011). However, the impacts of climate change are, of course, not restricted only to perishable items. In another very different example, steelmakers are now passing along their price increases by adding stiff clauses in contracts, allowing them to raise prices automatically (Matthews, 2011).

Service-Oriented Businesses

The logistics and/or physical flow considerations of the previous paragraph do not apply in general to the service sector. Services are highly dependent on the human resource input factor.

Those services are increasingly being performed over larger distances with telecommunications infrastructure fuelling offshoring, particularly during the past decades (Jahns, Hartmann, and Bals, 2006). What is crucial for such offshoring is the assurance that telecommunications connectivity of the service employees in the varying locations will be maintained.

In some instances, it is necessary that services are performed at a specific location (e.g., a facility management company offering cleaning services needs to be able to get to the buildings it is servicing). This need is driven by the inseparability of service and customers for some services (Zeithaml, Parasuraman, and Berry, 1985). The discussion to follow applies to situations in which the location of the employees performing the service and the location of the customers can be separated, such as in accounting, controlling, or call center services, which all heavily rely on broadband.

Here, disruption implies that the connectivity between the service-providing and service-receiving entities is interrupted. This could be the case for scenarios in which people cannot reach the facilities (e.g., due to flooding) or where the telecommunications infrastructure is destroyed (e.g., due to a thunderstorm damaging cables above ground or server farms).

Mitigation Strategies against Supply Disruption

Production-Oriented Businesses

It is not a simple task for suppliers to react upon short notice, so long and mid-term planning pay off in supply-disruption scenarios. In a major-scale event, such as the recent earthquake and tsunami in Japan, unaffected suppliers are in heavy demand. Although the supply chain managers were finding suppliers able to fill in the gaps made by the crisis in Japan, some of these suppliers started giving priority to their long-term customers (Mattioli, 2011). Such possibilities lead to a greater need to hedge against higher natural resource prices and the need for further attention to supply risk management (e.g., through scenario planning and contingency plans).

Two mitigation strategies are input reduction and developing the capability to use alternative resources thus reducing dependence on particular resources. Input reduction implies finding better designs that reduce the actual amount of material used. In the long term this approach could also include building recycling facilities to create a closed-loop process (i.e., the well-known cradle-to-cradle concept that will be discussed in more detail below). On the alternative resourcing side, the challenge is to find alternative inputs for equivalent products. An example would be the production of "plastic" cutlery and bags from corn starch, a truly multiple-sourced (and renewable) input in place of oil as an input with its increasing prices in an oligopolistic market. Additional strategies include localizing production and reducing variability of costs by relying on contract manufacturers with high resilience for at least part of production (Sirkin, 2011). For example, Nissan is planning to localize sourcing in the Americas to support growth in that region (Lariviere, 2011).

A concept that has gained increasing recognition in the area of using alternative resources is the "cradle-to-cradle" concept (Braungart and McDonough, 2009). This concept is contrasted with cradle-to-grave thinking in which products are designed for only single-life customer use. Cradle-to-grave means that raw materials go to parts, then go to the product, and then, after use, may be recycled (in a way that downgrades their characteristics) in one or a few recycles (i.e., downcycles), and then ultimately go to a landfill or incinerator. The alternative concept of cradle-to-cradle demands a design that takes into account the need for the product to cater toward customer needs, but also takes into account the effects it has during its entire life (e.g., in terms of toxic waste or emissions, use of water, et cetera) and if it lends itself to equivalently high-quality uses after its current design purpose is fulfilled. The overall idea is shown in Figure 5.1.

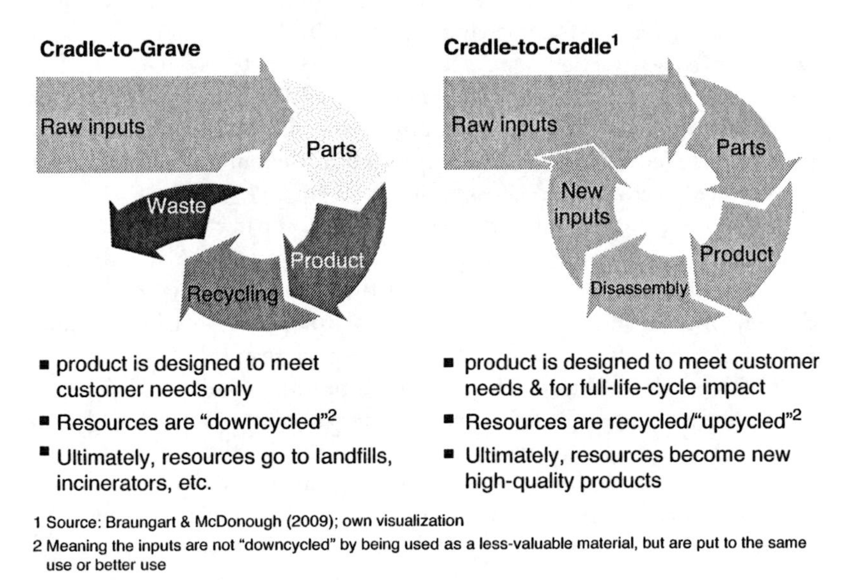

Figure 5.1 Cardle-to-Grave contrasted with Cradle-to-Cradle.

The key challenge in cradle-to-cradle design is creating technological or biological metabolism (Braungart and McDonough, 2009) in such a way that the next step in a product's life cycle is either to return to being used as high-quality technological inputs (not being "downcycled," but really "recycled" or "upcycled") or to go back to natural use as biological components for the production of other products (e.g., become nutritious, nontoxic fertilizer on a field). Braungart and McDonough are quite explicit in stating that these two metabolisms should not be mixed if avoidable during design, so that reassembly later on into one or the other cycles is easier.

When thinking about this technological metabolism from procurement, industry, or national perspective, the possibilities for decreasing dependency on complex supply chains is astonishing. By harvesting the resources "wasted" at the moment in nearby locations (e.g., in landfills), resource dependency across multiple-step supply chains reaching back to potentially disaster-prone (or politically unstable) countries of origin would be reduced. This process of filtering and leveraging existing resources is already sometimes referred to as "urban mining" (Mocker et al., 2009).

A recent example of finding creative ways to establish such "closed-loop" processes in Germany is a new technology by Currenta (a service firm of Bayer AG). Currenta has developed a process that leads to recycling yields for metals from scrap (such as cables, computers, et cetera) of 90–99 percent, based on a new approach with special ovens. The yield of conventional approaches is approximately 80 percent. In this new process, biological residuals are carefully burned while toxic waste is collected. The nonbiological materials remain and are thus available for reuse. The energy of the process is used for production processes at the production sites where the ovens are located, so the energy freed up in the burning process is not wasted (Leidinger and Bornewasser, 2010).

In another example of alternative resourcing, companies in Germany and many other countries are increasingly looking to increase the share of renewable energy they purchase, particularly in light of several anticipated changes in state and federal policies. These companies look to potential savings, positive reputation impact, and hedging against risk exposure to electricity price volatility and fuel supply disruptions (Smith, 2011).

In terms of design, an environmental perspective on product and process design is an extension of a quality management focus and directly affects the impact on natural resources and the environment. Effective design would incorporate such ideas as: (a) design for ease of disassembly, (b) design for disposability that will not have negative effects on the environment, (c) design to eliminate harmful processes in manufacturing, (d) design for ease of distribution and return, (e) elimination of many or all hazardous materials used,

(f) design for durability and reliability, and, of course, (g) design for customer success (Zsidisin and Siferd, 2001).

Service-Oriented Businesses

Mitigation strategies in service operations focus on being able to relocate quickly or to distribute the company's geographic footprint. Backup solutions may also need to be installed, whether they be completely up and running facilities ("hot backup") or "cold backup" such as facilities that are equipped, but not manned until an emergency occurs. Another cold backup approach involves arranging access to an additional flexible workforce, such as time-based teleworkers without permanent contracts who can be reached on very short notice.

There is also increasing pressure to diversify a company's global footprint by having multiple locations—so only one or few are affected by one temporary, regional natural disaster or political turmoil due to resource conflicts (e.g., by increasing food prices). This demand is converging with an ever-ready supply market, for example, such primary service offshoring locations as India, The Philippines, Russia, and Latin America. This pressure is, of course, combined with economic and other motives for outsourcing and offshoring such as "follow-the-sun," cost, talent, and/or flexibility. As was discussed in the context of production-oriented operations, diversification of footprint is a basic risk mitigation strategy. In the context of services rather than physical products, this basic strategy applies to human resources rather than fixed assets at a certain location.

It will be interesting to see how different industries react to this challenge, with diversifying the geographical footprint (with intracity, intercity, and/or intercountry backups) as one possibility, and with trying to localize as much as possible being another potential strategy. The choice of approach is likely to depend on the industry. For example, the more physical disruption that is possible in a natural resource or goods supply chain, the closer geographically the supply chain partners would want to be.

The Role of Procurement in Facilitating Sustainable Business

So far, the discussion has focused on the risk side of the implications of supply chain disruptions and which mitigation strategies can be taken. This focus is consistent with the notion that the sourcing activity in a sustainable business will be, in its simplest and traditional interpretation, to ensure supply security to maintain the company's ability to produce and distribute its products and/or services. To do so, risk within the supply chain is reduced.

These efforts need to follow a structured process. A general framework for sustainable supply chains has been proposed by Brammer, Hoejmose, and Millington (2011, p. 10), specifying the following four basic phases:

1. Code of Conduct: Used to set expectations of required conduct throughout the supply chain;
2. Certification: Used as a screening device in supplier selection and development;
3. Selection: Used as the primary process for reducing supply risks; and
4. Monitoring/auditing: Used to ensure compliance with expectations.

First of all, the code outlines expectations about supply chain behaviors. Then, certification is used as a screening device in supplier selection and development. Selection is performed with a clear focus on supply risk reduction. Finally, the loop is closed by monitoring and auditing the suppliers to ensure that they comply with the established expectations (Brammer, Hoejmose, and Millington, 2011). As highlighted by Brammer, Hoejmose, and Millington, these practices are the most prevalent ones in the literature for building supply chains and they reflect a "command and control" approach to supply chain management, since the lead buying firm is dictating the majority of the rules and processes.

To address shortcomings Brammer, Hoejmose, and Millington found in the simple model described above, the authors developed an extended model incorporating consultation, development, and learning. They state that the focal firm's power, relationships, resources, and needs will determine if it is in a position to pursue "next level" practice. Specifically, they now describe the four phases as including the following (Brammer, Hoejmose, and Millington, 2011, p. 11):

1. Create meaningful expectations: Environmental screening; code of conduct;
2. Confirm suppliers agree upon targets: Measurement, certification, and development of KPIs; supplier selection;
3. Measure supply chain performance: Monitoring; auditing; development; and
4. Evaluate and improve: data capture; supplier evaluation; learning.

Following this line of thinking, it becomes central that sustainable procurement practices are part of a coherent approach across the supply chain and not just an isolated auditing of some suppliers and a lighthouse project here and there.

Brammer, Hoejmose, and Millington's work calls attention to the question: What opportunities in a larger context can be discovered and exploited through procurement in the changing business environment? To address this question, it will be useful to explore what "sustainable business" means now and what it will mean in the future.

The interpretation of what "sustainable business" means will probably move beyond a carbon-only discussion, for example, and more toward the UN definition (World Commission on Environment and Development, 1987) and toward the definitions used in this volume to include the ecosystem and social dimensions. Of course, many companies (such as Patagonia or Newman's Own) are trying to operate on such dimensions (Epstein, 2008; Peredo and McLean, 2006, to name a few). One aspect of this expanded viewpoint is that there will still be CSR opportunities for businesses and their suppliers alike to differentiate themselves from competitors in image but also in tangible new products, creating and developing new niches, and whole new market segments. The challenge will be to align this effort with the suppliers, so sustainable practices can be ensured along multiple tiers of the supply chain. To do so, new models of coordination and cooperation will be needed.

As a study in the German automotive industry shows, a variety of standards are used. A comparison of the various standards that are currently used in the automotive industry, a very production-related setting (Beske, Koplin, and Seuring, 2008), reveals that these standards tend to do a poor job of covering more than one dimension (e.g., environmental, social, process, et cetera). A standard measure with multidimensional coverage is still needed. Another important aspect to keep in mind is that these standards have assumptions about the context in which they are applied and that this context may look very different in developed versus emerging countries, where infrastructures (e.g., waste disposal) may not be in place at all (Lund-Thomsen, 2008). Both multidimensional criteria and consideration of the operating context of the company and its suppliers are important for the above-mentioned coordination and cooperation.

In a recent study in which 378 senior executives, including 86 from the United States, were surveyed, over 50 percent stated that their company had adopted a formal sustainability strategy and an additional 31 percent plan or expect to adopt one (KPMG, 2011). The main drivers were to enhance brand reputation, regulatory, or legal compliance, and reduce costs. These responses suggest how much sustainability has climbed up the corporate agenda.

Transparency is also becoming a core issue, offering an opportunity to create consumer confidence across whole industries (and to avoid unnecessary duplication and competition among differing standards). Translating this opportunity into a goal, some companies have started to form sustainability indices, such as Walmart's and Patagonia's Sustainable Apparel Coalition, which is made up of more than 30 annual dues-paying companies and environmental organizations that are now working together to develop an industry-wide supply chain index that will measure water and energy use, greenhouse gas emissions, waste, labor practices, and other factors. The index will combine elements of the OIA Index, Nike Considered Index, and parts of Walmart's Supplier Sustainability Assessment Tool (Schwartz, 2011). Such indicators provide the procurement function guidance about what is expected of the supply base and can become requirements along the supply chain.

Sustainability is increasingly becoming an imperative, not just a "nice to have" ethical box to check off in the annual report (Bettridge, 2010). The rules for business are changing as suppliers are being affected by more restrictions, the changing of customer attitudes, and governmental regulations.

As suppliers move to comply with new environmental regulations, they may face difficulties (a risk management interpretation) or they may react to new regulations by developing new products themselves. The increasing importance of developing "sustainable products" as a source of competitive advantage in growing markets emphasizes the role of suppliers as sources of innovation. It is also necessary to collaborate with suppliers to help them cope with climate change, to avoid disruptions, and to ensure that they are following sustainable practices so one's own supply chain can warrant the label "sustainable." These latter considerations apply particularly to manufacturing businesses, though service-providing businesses also have huge opportunities for carbon reductions in particular (e.g., videoconferencing and telecommuting) in their operations, as well as leveraging opportunities to be socially responsible.

Monitoring leading customers seeking sustainable products/services (e.g., those with the smallest possible resource footprint and/or incorporating the cradle-to-cradle concept) and leveraging them to start moving up the learning curve becomes a vital opportunity for businesses in opening up new market segments or completely new markets. These market trends need to be analyzed in conjunction with regulatory requirements now in place or in the making. But this ability to sense downstream opportunities is a core task of the marketing function and encourages a shift of focus upstream to the procurement area. The ability to monitor and react to changing regulations is a key ability for a firm, and also a core task of legal departments. The collaboration of different departments including marketing, procurement, and the legal department provides one indication that successfully addressing these opportunities will be a truly cross-functional challenge.

A potential place for procurement to start involves taking care of some of the low-hanging fruit, which can be found in sourcing and transportation for a lot of companies (Blue, 2011). This perspective was addressed in a study conducted by Lefevre, Pelle, Abedi, Martinez, and Thaler (2010), in which product categories were classified in terms of cost-reduction potential and ease of implementation. Packaging, transportation, and energy-using products offer particularly attractive opportunities because they have high cost-reduction potential and low implementation costs, placing them in the "hot category." The three "cold categories" are characterized by (1) high cost-reduction potential with low ease of implementation (such as energy production, primary raw materials, industrial equipment, building, et cetera); (2) low cost-reduction potential with low ease of implementation (such as secondary war materials and monopolistic products); and (3) low cost-reduction potential with high ease of implementation (such as nonregular purchase of goods and services, low cost products/services).

In the packaging context, sustainable business practices can yield significant savings for companies, such as PepsiCo and Walmart, as price volatility in plastic packaging, fuel, and food increase the sense of urgency for their initiatives and ultimately might have a real impact on shareholder value in the next decade (Stanford, 2011). For example, Walmart's Japanese Seiyu chain converted packages for its private-label fresh-cut fruit and salads from oil-based to corn-based plastic in 2009. This change reduced the packaging's weight by 25 percent and cost by 13 percent, resulting in savings of over $195,000 a year (Stanford, 2011).

The 2010 study by Lefevre et al. highlighted the return on investment of sustainable procurement practices on the criteria of cost reduction, risk mitigation, and revenue growth. The cost-reduction initiatives they studied yielded up to a six times payback on procurement program costs. For revenue growth (induced by innovation of ecofriendly products/services, price premiums, or additional income from recycling activities) there was up to 58 times payback (Lefevre et al., 2010).

In the effort to accommodate corporate sustainability goals, it can be a great help to be explicit in identifying and discussing the priorities of the effort. Here, Braungart and McDonough (2009, p. 150) recommend a discussion along three dimensions, similar to the UN triangle of economic, environment, and social. Using their triangle in which the social dimension is termed "equity," a discussion can be triggered on how to position the company's current and upcoming projects. For example, if the economic dimension is the primary trigger, cost savings or revenue growth opportunities would be prioritized. If the target is maximizing equity (e.g., fair treatment

Spot risks

- Help identify risks
    - Input disruption (-> find other sourcing options)
    - Transportation disruption (-> geographic footprint and inventory decisions)
    - Production disruption (-> geographic footprint and contract manufacturing options)

Spot opportunities

- Help identify suppliers with
    - Alternative or fewer inputs
    - A product (component) or process facilitating a new business model
    - More efficient transportation
    - Recycling and reusing options to close the cradle to cradle loop

Figure 5.2 Procurement as risk and opportunity spotter.

of employees, equal pay for men and women, et cetera), the goal can be to improve these situations, perhaps with no profit improvement at all.

Figure 5.2 characterizes the aspects of managing risks and leveraging opportunities that have guided this chapter thus far. On the risk side, input disruption (e.g., risks arising at the stage of raw material suppliers) has to be kept in mind, as well as how transportation and production (or service delivery) need to be secured. On the opportunity side, it is important to keep in mind possibilities for alternative inputs, products, or processes facilitating new business models (e.g., innovations allowing green products), ensuring more efficient transportation, and finally closing the loop by recycling or reusing.

Min and Galle (2001, p. 1228) define recycling and reusing clearly. "Recycling" comprises the key activities of collecting, separating, processing, and remanufacturing (including organic materials). Solid waste is the focus of recycling. The objective is to use the material to be recycled in new products. In contrast, "reusing" involves sorting, recirculating, redistributing, refurbishing, and repairing without remanufacturing, creating second-hand products that have the same form and function as new products.

Min and Galle (2001) note that recycling and reusing are relatively more applicable at different stages in the supply chain. Recycling is more involved in the manufacturing stage, while reusing is more concerned with the distribution stage. Therefore, procurement interfaces for these two activities involve different functions. The next section of this chapter focuses on the cross-functional nature of identifying and implementing sustainable procurement practices.

Looking Ahead: A Cross-Functional Challenge

Cross-functional cooperation is key for reducing risks and leveraging opportunities. In most cases procurement, in isolation, cannot induce radical change. The secret of success for sustainable procurement is developing a deep analytic understanding of the organization's overall strategy, including the roles of all functional areas and their alignment. Achieving shared understandings and cross-functional collaboration requires commitment and effort.

Common definitions of sustainable procurement and sustainable supply chain management call attention to the importance of cross-functionality. Narasimhan and Carter (1998, p. 6) defined environmental supply chain management as "the purchasing function's involvement in activities that include reduction, recycling, reuse, and the substitution of materials." This definition was extended by Zsidisin and Siferd (2001, p. 69) to read: "Environmental Supply Chain Management (ESCM) for an individual firm is the set of supply chain management policies held, actions taken, and relationships formed in response to concern related to the natural environment with regard to the design, acquisition, production, distribution, use, reuse, and disposal of the firm's goods and services." This latter definition highlights the high cross-functional nature of this challenge.

Figure 5.3, which employs Porter's (1985) value chain perspective, illustrates the issues that need to be taken up and clarified with the other functions. Figure 5.3 adds examples of the opportunity-spotting aspect mentioned earlier and suggests how procurement can add value to the individual functions. The challenge lies in agreeing upon common objectives and pursuing them in a true cross-functional effort to achieve the best results.

An example of successes in cross-functional collaboration is offered by Dupont's recent developments in sustainable packaging that accelerate the growth of PLA polymers, decrease flexible packaging waste and expenditure, and improve the convenience of peelable lids (Quality Assurance and Food Safety Magazine, 2011). In such efforts collaboration between logistics and procurement is the key. Moreover, internal customers need to agree to a very high level of standardization. In diversified companies this step may be a challenge, especially as marketing functions need to come to a consensus based on estimations of how changes might affect their product portfolio sales.

Procurement can also help to suggest how to integrate supplier packaging innovations. As this task primarily involves design and product characteristics, the collaboration challenge tends toward R&D (or "technology development" as spelled out in the original value chain). For example,

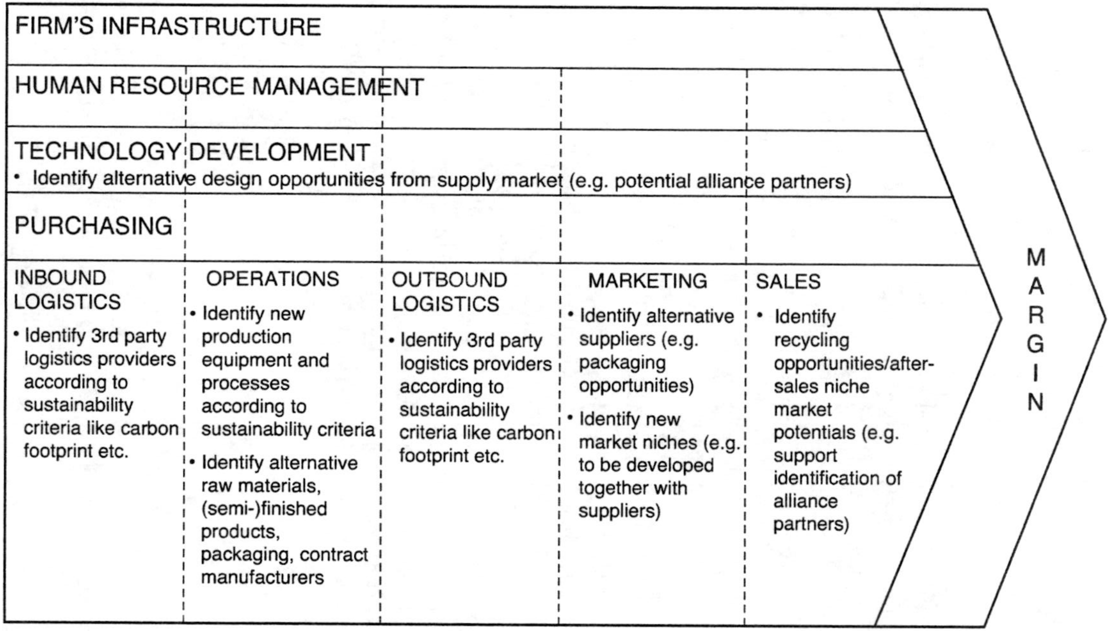

FIRM'S INFRASTRUCTURE

HUMAN RESOURCE MANAGEMENT

TECHNOLOGY DEVELOPMENT

- Identify alternative design opportunities from supply market (e.g. potential alliance partners)

PURCHASING

INBOUND LOGISTICS

- Identify 3rd party logistics providers according to sustainability criteria like carbon footprint etc.

OPERATIONS

- Identify new production equipment and processes according to sustainability criteria
- Identify alternative raw materials, (semi-)finished products, packaging, contract manufacturers

OUTBOUND LOGISTICS

- Identify 3rd party logistics providers according to sustainability criteria like carbon footprint etc.

MARKETING

- Identify alternative suppliers (e.g. packaging opportunities)
- Identify new market niches (e.g. to be developed together with suppliers)

SALES

- Identify recycling opportunities/after-sales niche market potentials (e.g. support identification of alliance partners)

MARGIN

Figure 5.3 Procurement opportunities from a cross-functional perspective.

Source: Adapted from Porter, 1985.

Kraft's R&D and procurement functions developed a jointly owned supplier assessment process that enabled the company to access critical external innovation capabilities without sacrificing cost considerations. This collaboration resulted in such tangible benefits as a differentiated packaging solution and overall reduced time to market, as well as significantly improved revenues (Corporate Executive Board, 2008).

These cross-functional collaboration considerations call attention to facilitators and barriers to the involvement of the procurement function in opportunity-spotting generally. Typical barriers include (a) lack of awareness, (b) lack of skills, (c) lack of motivation, and (d) lack of opportunity. These barriers also highlight the angles from which to tackle lack of involvement (e.g., to create awareness by finding corporate ambassadors for procurement and issuing clear guidelines and policies in media and formats that are widely read). Or, for example, tackling lack of skills by building up skills and knowledge for the subject matter area procurement employees are meant to support (Bals, Hartmann, and Ritter, 2009).

Conclusion

Throughout this chapter two basic questions were addressed:

1. What implications does climate change have for the procurement function and for supply chains?
2. What implications will climate disruptions have for what it means to be a "sustainable business" and for managing the procurement process in such businesses?

For the first question, the implications of supply disruptions (induced by weather anomalies directly, such as floods, or indirectly, such as political upheaval due to food shortages caused by exceptional humidity or aridity) were laid out with mitigation strategies. For the second question, the implications of a more holistic sustainability interpretation were discussed, including moving more and more beyond a pure carbon-reduction focus to encompassing all three dimensions of the sustainability triangle, i.e., environment (such as avoiding biodiversity loss), social (such as achieving fair working conditions), and economic (such as keeping companies economically viable). The importance of being clear on where a company is located in the environmental-social-economic triangle was emphasized as an important precondition for having procurement facilitate these corporate or functional goals. Moreover, the importance of cross-functional collaboration was highlighted as a key success factor.

By analyzing these questions and showing examples of the latest developments in this area, this chapter sheds light on the opportunities and risks arising for the procurement function. As was stated, it makes a lot of sense to differentiate more production-oriented and more service-oriented settings when analyzing which risks could occur and also when evaluating which mitigation strategies might be taken.

On the risk side, procurement can create value for companies by ensuring business continuity and providing early warning information. In the production setting, the necessity for doing so is aggravated if the company is operating on a low-stock model and/or perishable goods are involved. Here, the procurement function can fulfill the role of a risk-reducer.

Considering the opportunity side, forward-looking procurement has a chance to act as a driver and central facilitator of corporate innovation, ensuring that internal sustainability initiatives are successful and/or that new external products/service offerings can be used to embed sustainability in the company's offerings portfolio. As was highlighted, innovative approaches that have to be kept in mind here are input minimization design opportunities as well as ideas such as alternative resource use. For these latter approaches, the cradle-to-cradle concept was explained and identified as an important vehicle for helping to establish the procurement function as an opportunity-spotter.

Successful cross-functional cooperation between procurement and other functions was spelled out in the final section of this chapter as being central to achieve maximum impact in risk reduction, as well as on opportunity identification and exploitation. In most companies procurement in isolation will not achieve radical changes in sourcing decisions. It will have to align itself with other functions to be successful. Therefore, it is key for procurement professionals to show how they can add value to the organization and to approach other departments in the firm with clear ideas and a convincing competence portfolio.

Climate change impacts are already having effects on business, and procurement professionals currently have a window of opportunity to create a lot of value for companies. If sustainability is taken up as a shared objective, it has true potential to be a lever that can take the procurement function to the next level. A shared sustainability objective could be a great opportunity to overcome the cross-functional barriers that still exist today at so many companies and could empower procurement professionals to fulfill an integral role as risk-reducers and business opportunity-spotters heading into the next decade.

References

Bals, L., E. Hartmann, and T. Ritter. 2009. Barriers of purchasing involvement in marketing service procurement. *Industrial Marketing Management* 38(8): 892–902.

Barriopedro, D., E. M. Fischer, J. Luterbacher, R. M. Trigo, and R. García-Herrera. 2011, March 17. The hot summer of 2010: Redrawing the temperature record map of Europe. *Science*. doi: 10.1126/science.1201224.

Beske, P., J. Koplin, and S. Seuring. 2008. The use of environmental and social standards by German first-tier suppliers of the Volkswagen AG. *Corporate Social Responsibility and Environmental Management* 15, 63–75.

Bettridge, N. 2010, December 8. Sustainability and the CFO. *Finance Director Europe*. http://www.the-financedirector.com/features/feature104156/ (Accessed July 26, 2011).

Blue, G. 2011, March 23. How to build sustainability into your supply chain. *Inc. Magazine*. http://www.inc.com/guides/201103/how-to-build-sustainability-into-your-supply-chain.html (Accessed July 26, 2011).

Brammer, S., S. Hoejmose, and A. Millington. 2011. *Managing sustainable global supply chains*. London, ON: Network for Business Sustainability. http://www.nbs.net/wp-content/uploads/Supply-Chain-Report.pdf (Accessed July 26, 2011).

Braungart, M., and W. McDonough. 2009. *Cradle to cradle: Remaking the way we make things*. London: Vintage.

Campbell, V. 2011, March. A superstorm in the forecast. *Risk Management Magazine*. http://www.rmmag.com/MGTemplate.cfm?Section=RMMagazine&NavMenuID=128&template=/Magazine/DisplayMagazines.cfm&IssueID=353&AID=4272&Volume=58&ShowArticle=1 (Accessed July 31, 2011).

Corporate Executive Board, Procurement Strategy Council. 2008. *Enhancing alignment for growth: Maximize supplier innovation value capture through R&D-Procurement collaboration*. Washington, D.C.: Corporate Executive Board.

Epstein, M. 2008. *Making sustainability work: Best practices in managing and measuring corporate social, environmental, and economic impacts*. San Francisco: Berret-Koehler.

Jahns, C., E. Hartmann, and L. Bals. 2006. Offshoring: Dimensions and diffusion of a new business concept. *Journal of Purchasing and Supply Management* 12(4): 218–231.

KPMG. 2011. *Corporate sustainability: A progress report*. Amstelveen, Netherlands: KPMG. http://www.kpmg.com/Global/en/IssuesAndInsights/ArticlesPublications/Pages/corporate-sustainability.aspx (Accessed July 26, 2011).

Lariviere, M. 2011, April 25. Layers of supply chain risk. *The Operations Room*. http://operationsroom.wordpress.com/2011/04/25/layers-of-supply-chain-risk/ (Accessed July 26, 2011).

Lefevre, C., D. Pelle, S. Abedi, R. Martinez, and P. F. Thaler. 2010. Value of sustainable procurement practices: A quantitative analysis of value drivers associated with sustainable procurement practices. London: Pricewaterhouse Coopers.

Leidinger, W., and U. Bornewasser. 2010. Edelmetalle aus dem Drehrohr. *Umweltmagazin*, 12/2010, 36–37.
Lobell, D., W. Schlenker, and J. Costa-Roberts. 2011, May 5. Climate trends and global crop production since 1980. *Science*. doi:10.1126/science.1204531.
Lund-Thomsen, P. 2008. The global sourcing and codes of conduct debate: Five myths and five recommendations. *Development and Change* 39(6): 1005–1018.
Matthews, R. 2011, May 3. Steelmakers turn to price triggers as iron-ore costs rise. *The Wall Street Journal*. http://online.wsj.com/article/SB10001424052748704569404576299490008326696.html (Accessed July 26, 2011).
Mattioli, D. 2011, April 4. Disaster highlights component firms. *The Wall Street Journal*. http://online.wsj.com/article/SB100014240527487036967045762226922972551676.html (Accessed July 26, 2011).
Min, H., and W. Galle. 2001. Green purchasing practices of US firms. *International Journal of Operations and Production Management* 21(9/10): 1222–1238.
Mocker, M., K. Fricke, I. Löh, M. Franke, T. Bahr, K. Münnich, and M. Faulstich. 2009. Urban mining—Rohstoffe der zukunft. *Müll und Abfall* 10: 492–501.
Narasimhan, R., and J. Carter. 1998. *Environmental supply chain management*. Tempe, AZ: Center for Advanced Purchasing Studies.
Peredo, A., and M. McLean. 2006. Social entrepreneurship: A critical review of the concept. *Journal of World Business* 41(1): 56–65.
Porter, M. 1985. *Competitive advantage: Creating and sustaining superior performance*. New York: The Free Press.
Quality Assurance and Food Safety Magazine. 2011, March 31. DuPont introduces new developments in sustainable packaging. http://www.qualityassurancemag.com/Article.aspx?article_id=114780 (Accessed July 26, 2011).
Reddy, S. 2011, April 5. Unwanted new item on menu: Higher prices. *Wall Street Journal*. http://online.wsj.com/article/SB10001424052748703806304576243203052636840.html?mod=googlenews_wsj (Accessed July 26, 2011).
Schneider, H. 2011, March 14. Higher food prices may be here to stay. *The Washington Post*. http://www.washingtonpost.com/business/economy/higher-food-prices-may-be-here-to-stay/2011/03/10/AByYO3V_story.html?wpisrc=nl_wonk (Accessed July 26, 2011).
Schwartz, A. 2011, February 25. Patagonia, Adidas, Walmart team up on sustainable apparel coalition. *Fast Company*. http://www.fastcompany.com/1731780/patagonia-hm-walmart-team-up-on-a-sustainable-apparel-index (Accessed July 26, 2011).
Sirkin, H. 2011, March 28. How to prepare your supply chain for the unthinkable. *Harvard Business Review Blog*. http://blogs.hbr.org/cs/2011/03/why_are_supply_chains_eternall.html (Accessed July 26, 2011).
Smith, B. 2011, April 6. Growing your bottom line through energy management. *Industry Week*. http://www.industryweek.com/articles/growing_your_bottom_line_through_energy_management_24265.aspx (Accessed July 26, 2011).
Stanford, S. 2011, March 31. Why sustainability is winning over CEOs. *Businessweek*. http://www.businessweek.com/magazine/content/11_15/b4223025579541.htm (Accessed July 26, 2011).

World Commission on Environment and Development. 1987. *Our common future.* Oxford, UK: Oxford University Press.

Zeithaml, V., A. Parasuraman, and L. Berry. 1985. Problems and strategies in services marketing. *Journal of Marketing* 49: 33–46.

Zsidisin, G., and S. Siferd. 2001. Environmental purchasing: A framework for theory development. *European Journal of Purchasing and Supply Management* 7: 61–73.

CHAPTER 6

Business and Climate Change Adaptation: Contributions to Climate Change Governance

Nicole Kranz

Introduction

The relationship between business and climate change is subject to an increasingly intensive discussion in the environmental community, academia, business, and even the public at large. Businesses are undoubtedly important factors when it comes to approaching the phenomenon of climate change and they are often considered the main culprits for environmental degradation, and thus, climate change. Much of this perception focuses on climate-relevant emissions from production processes (WBCSD, 2009). While this concern has led to widespread criticism of business practices, there is also a growing awareness that without contributions from and fundamental changes within the business community, it will not be possible to address the challenges emerging from global climate change. There is a need to consider carefully the roles that businesses can play to address these issues, and what motivations might apply in driving business behavior. These considerations are also supported by analyses conducted by Van Zeijl-Rozema, Cörvers, Kemp, and Martens (2008) on the mode of governance for sustainable development, which is described as complex but also as necessarily inclusive of all potentially relevant actors, as states alone might not be in a position to address these paramount challenges. According to Liu et al. (2007), climate change epitomizes these types of complex, nonlinear interactions between social and natural systems that can be labelled "wicked problems." Wicked problems are ones that need to be addressed in new and creative ways, many of which will be dramatically different from past "business as usual" ways of "solving," or at least attempting to solve, less complex problems.

If we look to the current academic literature addressing the relationship between business and climate change, much focus is placed on strategic orientation at the firm level, with a strong emphasis on the mitigation of climate-relevant emissions as well as those changes necessary within a business organization to cope with the potential impacts of climate change on operations (Linnenluecke and Griffiths, 2010; Pinkse and Kolk, 2010; Winn, Kirchgeorg, Griffiths, et al., 2010). Climate change in these cases is often framed as a strategic risk for businesses and proposed solutions are based on derivatives of business strategies, which are adapted to the climate change challenges. These solutions are for the most part directed at rendering a more resilient business strategy and thus at protecting or even expanding the income base for the firm as a strategic entity.

This chapter takes a slightly different perspective by focusing on business adaptation to climate-induced environmental changes. It is based on the premise that environmental change affects businesses and the surrounding communities alike and thus it places a strong emphasis on the interaction between these two groups. The chapter covers firm-internal adaptation measures as well as those measures that originate from companies but have strong external implications. This focus covers the more traditionally addressed internal changes and consequences and the less frequently addressed external company actions that involve interaction with surrounding communities and may yield, or not yield, improving overall adaptive capacity of both company and community. In this regard, the focus is on actions that may transcend the immediate sphere of firm influence and contribute to overall climate change governance (Hamann, 2010). The chapter also investigates potential drivers that can motivate businesses to adopt sustainability-oriented strategies or discourage them from doing so. These drivers are a combination of firm-internal factors such as an increased awareness of business vulnerability to climate change, and external drivers, such as regulatory incentives. In combining these two drivers of sustainable business practices, the chapter aims to merge the current debate on corporate responsibility with considerations of climate change governance that reflects incipient developments at the policy level (WBCSD and IUCN, 2010).

Methodology and Approach

In the sections that follow, necessary conditions for climate change adaptations are derived, using the example of sustainable water management as a proxy. Potential business strategies and responses for sustainable water management are then developed. Potential drivers influencing business behavior are also discussed. This discussion is used to establish initial assumptions that are then applied to a case situation to refine and "verify" the initial model and assumptions. This methodological approach is guided by concepts developed by Eisenhardt (1989) and George and Bennett (2005). Those concepts propose the use of iterative case study research to arrive at a more comprehensive and accurate picture of the interactions and relationships proposed by existing literature. The South African mining industry serves as the case study for testing some of the initial assumptions about the contribution businesses can make to climate change adaptation. In the last section of the chapter, preliminary findings are used to derive policy recommendations for decision-makers and the identification of potential further research activities.

Adaptation to Climate Change Through Sustainable Water Management

Adaptation measures can involve a wide range of activities, depending on the climate change impacts incurred. In this chapter, specific focus is placed on the impact of climate change on water management.

There is an abundance of evidence that suggests that water resources will be significantly impacted by climate change with wide-ranging further impacts on ecosystems and societies (Bates, Kundzewicz, Wu, et al., 2008). The expected impacts of climate change, such as increased precipitation on the one hand and droughts on the other, are expected to vary substantially, both in geographic location and in intensity. For example, some semiarid areas will experience a further decrease of water resources, while other regions will be exposed to a higher likelihood of flooding due to increased precipitation intensity.

In addition to these immediate impacts on water resources, further impacts can be expected for food security, the operation of water-related infrastructure, energy production, land management, public health, and nature conservation. Water resources are a crucial element in all of these policy areas and thus serve as a useful vehicle for considering climate change adaptation policies (Kranz, 2010). So, studying public and private approaches to sustainable water management as a means of addressing the impact of climate change on water resources, and thus on many other aspects of global sustainability, provides a useful way of exploring the potential contributions that businesses can make to address these and perhaps other challenges.

The technique of using the set of requirements for sustainable water governance as a point of departure for the ensuing analysis is based on the assumption that the characteristics of water resources necessitate a specific rule system. Such a set of rules has been described by Ostrom (1990) in her work on the appropriation and provision of common property resources. Based on this framework, a set of normative requirements can be derived that would provide for the sustainable governance of such complex systems as water (Dietz, Ostrom, and Stern, 2003).

Among the key factors the literature suggests might be needed for the sustainable governance of complex systems, such as water systems, are: information, participatory decision-making, infrastructure, preparedness for change, and integration. These factors are discussed in more detail under those headings.

Information: The provision of sound, trustworthy *information* is considered key in making decisions about complex environmental systems. In addition, such data need to be congruent with the scale of the problem addressed (Young, 2002) and to cater to the needs of the decision-makers. In accounting for the uncertainties inherent to the management of natural systems, the information provided needs to give an estimation of the uncertainties involved as well as to allow for the assessment of trade-offs encountered across multiple scales.

Participatory decision-making: Since the management of natural resources always bears the potential for conflict among different stakeholders, adaptive governance systems must be geared toward avoiding and addressing potential or actual conflicts. Doing so requires establishing mechanisms for potentially conflicting parties to *participate in decision-making*, thus creating arenas for learning and change (Pahl-Wostl, 2008). This assumption for this chapter (and also a conclusion by Pahl-Wostl) is corroborated by the view on water governance suggested by Turton et al. (2007), who propose the use of a trialogue model to reveal and explore the interfaces between different actor groups that need to be involved in adaptive water management approaches. Turton et al. argue that government, society, and the scientific community need to engage in a mutual dialogue if there is to be a transition to more adaptive water management. They describe the relationship between government and society as "an unwritten, hydro-social contract, incorporating the norms and values of society that structure the relationships between key stakeholders" (Turton et al., 2007:21). They point to the interfaces between these different actors as important loci where dialogue and learning take place and where common values are developed, that then in turn affect the resilience and robustness of the water system under management.

The aspect that emerges most strongly in both cases discussed in this chapter, is that of learning, or rather social learning in the context of reflexive governance. Learning and reflexivity are achieved through dialogue and interaction of all stakeholders involved (Pahl-Wostl, 2008; Turton et al., 2007).

Infrastructure: As indicated before, *infrastructure* constitutes a further key component in the management of water resources. According to Dietz, Ostrom, and Stern (2003), the role of infrastructure can be conceptualized in two ways. On the one hand, infrastructure supports the exploitation of the resources and thus the use of infrastructure needs to be subjected to careful planning. On the other hand, infrastructure can be used to protect natural resources, help provide equitable access to these resources, enable the monitoring of human impact and assist with the generation of information for planning purposes.

Preparedness for change: Dietz, Ostrom, and Stern furthermore advocate certain *preparedness for change* of the institutions established for managing the natural resources in question. This concept draws on some of the previous factors and represents the principal lesson of adaptive management research (Gunderson and Holling, 2001). More recently, in the face of global environmental change, research has addressed the necessity to cope with present and future changes and uncertainties. This need for being ready to be changed and to bring about changes is assumed to be a key component of sustainable water management systems (Global Water Partnership—TAC, 2004).

Integration: The Global Water Partnership, an international network of experts in the field of water management, promotes *integration* as one of the key attributes of water governance systems (Global Water Partnership—TAC, 2004). Integration would include the linkages between macroeconomic policies and water development, water management, and water use. It would also comprise the integration across different sectors (e.g., industry, agriculture, and households) and the integration of decisions made at the local and river-basin level with broader national objectives. Integration as a guiding management paradigm for achieving sustainable water management is referred to as integrated water resources management (IWRM), and is broadly defined as" coordinated development and management of water, land, and related resources, in order to maximize the resultant economic and social welfare in an equitable manner without compromising the sustainability of vital ecosystems" (Global Water Partnership, —TAC, 2000). In 2002, this concept gained additional support at the World Summit on Sustainable Development in Johannesburg, where the goal was established for all countries to develop integrated water resources management plans by 2005 (Bullock, Cosgrove, Van der Hoek, et al., 2009). Consistent with work by Holling (1978), Pahl-Wostl and Sendzimir (2005) suggest that the concept of adaptive management can be used as a central management style for achieving effective IWRM processes.

Although these concepts are somewhat ill-defined, adaptive and integrated management serve as core frames of reference for sustainable water management. In addition, these management paradigms are augmented by a number of workable management principles.

The *precautionary principle* refers to careful *planning* and the use of scenarios in selecting water management measures. Measures should be chosen according to the no-regret-principle, and ideally, they can be reversed (O'Riordan and Cameron, 1994), thus supporting the requirement of preparedness and adaptive management.

The *polluter-pays-principle* points to the *responsibility* of those causing the pollution or deterioration of water resources for mitigating those impacts retroactively, whether through undertaking the clean-up or providing financial means for others to perform this task (Rogers, De Silva, and Bhatia, 2002).

The *principles of source reductions and resource minimization* in the first instance address the need to limit potential pollution sources, for example, through redesigning production processes or installing end-of-pipe clean-up mechanisms. Second, these principles advocate the reduction of the resources used, for example, through the increased efficiency of production processes (Molden, 2006; Umweltbundesamt, 2001). These principles also speak to the IWRM components of water resources assessment and careful planning for IWRM as well as efficiency in water use.

In reviewing the characteristics and requirements for sustainable water management outlined above, the overall governance style, echoing observations by Dietz, Ostrom, and Stern (2003), should be predictable, open, and enlightened in terms of a clear vision about water management. At the same time, *professional and well-capacitated institutions* would be required. These institutions would have sufficient human and financial resources available to them, know how to manage infrastructure and knowledge, have the ability to form partnerships, and at the same time allow for broad-based civil society participation and the involvement of other non-state actors in developing mutually agreed adaptive water management options. While successful strategies will be based on this pattern and logic (Dietz, Ostrom, and Stern, 2003), the actual type of the institutional set-up will eventually vary with the specific water management situation as well as other factors deriving from politics and culture in a country or region. A further differentiation needs to be introduced between institutions at the national, basin, and local level (Pegram, Orr, and Williams, 2009).

These attributes and requirements of sustainable water governance serve as a basis for identifying and evaluating business strategies and activities that can assist in addressing key water management challenges. Those possible business contributions are discussed next, followed by an investigation of motivations and key drivers defining corporate strategies.

Role of Businesses

The contributions to sustainable water management that businesses can make are to a large extent determined by corporate strategic options (and eventually choices), which can either take place within the firm's boundaries or beyond, that is, along the value chain and/or in interaction with other stakeholders (Hoffman, 2000).

Examples of sustainable activities taking place mainly within the firm's boundaries include increasing the efficiency of water usage, reducing water intake for production purposes as well as improving the quality of sewage released into the environment. Through these measures, firms cater to the principles of resource reduction and source minimization and thus relieve the overall pressure on the water ecosystem in terms of water usage and/or quality implications (Hoffman, 2000). Firms can also engage in establishing similar approaches with suppliers or in other parts of their supply chain.

Internal monitoring and water resources planning also support the careful use of water resources by businesses. Monitoring is also an important means to ensuring overall compliance with existing legislation. When businesses' monitoring data and water resource plans that apply to the watershed level of a particular region are shared with surrounding communities and discussed with other stakeholders, such information dissemination constitutes an important contribution to the water resources governance in that particular region (Dietz, Ostrom, and Stern, 2003). When business planning also caters to the precautionary principle, it aids in improving the preparedness to changes in water availability as well as other climate-change-induced impacts.

Further potential contributions of private actors derive from their involvement with water infrastructure planning, development, and financing. In cases where firms become involved with developing water infrastructure, doing so may not only benefit their own operations but also help meet community needs (Addams, Boccaletti, Kerlin, et al., 2009; Pegram and Schreiner, 2009). Related contributions pertain to capacity-building, e.g., in the operation and maintenance of water infrastructure. In addition, capacity-building activities provided by companies might also include capacity-raising measures to improve planning and monitoring skills for government institutions as well as community representatives (Pegram, Orr, and Williams, 2009). When firms become involved in shaping policy dialogues and awareness-raising activities, their doing so might lead to further learning and adaptation processes for local institutions, and for the businesses themselves. Through these activities, firms actively shape their governance environment, ideally helping to increase the adaptive capacity of surrounding communities through more stable water infrastructure, better administrative capacity, and better knowledge and planning. These types of contributions do not need to be restricted to companies and their immediate sphere of influence. Rather, they can occur at very different levels ranging from local municipalities to the basin levels, as well as in the national context (Loorbach, Van Bakel, Whiteman, et al., 2009).

Potential Motivational Patterns

Motivations and drivers for business behavior can be discussed as firm-internal factors on the one hand and institutional drivers on the other (Bansal, 2005). Internal drivers are those factors arising from the resource base of a firm, ranging from human resources to financial considerations. Natural resource constraints and the cost of resources are also included in this category (Hart, 1995). Human resources are influenced by employee motivation as well as the overall relevance of sustainability policies within a firm (Swanson, 2008). Other factors include the availability of resources (organizational slack) and the ability to manage capital and assets.

Turning to institutional drivers and following the work of Scott (2001), international sustainability norms (specifically in the areas of corporate responsibility, water management, climate change, and similar drivers) are distinguished from specific industry norms (Delmas and Toffel, 2004). Next, competitive and market drivers are considered, including those deriving from consumers (Smith, 2008), investors (King and Lenox, 2001), and external stakeholders such as NGOs, trade unions, and community groups.

Finally the influence of government actors is discussed, focussing on the different roles government actors take (Fox, Ward, and Howard, 2002). Possible roles of government actors range from mandating, which includes the strict monitoring and enforcement of regulations, to softer modes of interaction, including facilitating, partnering, and endorsing sustainability-focused initiatives. Specific attention is paid to the relevance of government capacity as a determining factor for government intervention (Schwartz, 2003).

The analysis is based on the premise that government actors at several levels need to possess appropriate capabilities to be able to take on various roles, such as regulator and/or partner, successfully. These capacities rely on the availability of appropriate financial and human resources, but also on the ability to engage a wide range of actors. The lack of these capacities on the part of government can act as a special kind of motivator for businesses. Perceived inability at the government level to address particular challenges may prompt businesses to engage in these functions (Kranz, 2010).

In the section that follows, the framework developed in the previous sections from research and study of the water resources management challenge is applied to the case of the mining industry in South Africa.

The Mining Industry in South Africa: An Indicative Case Study

The South African mining sector was chosen for exploring the assumptions about business contributions and drivers derived above because South Africa is an increasingly water-stressed country (Ashton and Hardwick, 2008) and mining has a significant impact on water resources.

South Africa is faced with water-quality deterioration, water scarcity, and the challenge of providing water services to an increasingly urbanized population. The nation is likely to be severely affected by continuing climatic changes (Mukheibir, 2008) and some industrial regions are likely to suffer most from decreasing water availability. As a result, South Africa encounters both typical water problems of the industrialized world and those of developing countries.

The South African mining sector has a significant impact on water resources due to pollution incidents, freshwater abstraction, and infrastructure developments that seek to assure water supply to the industry (Ashton, Love, Mahachi, et al., 2001). The industry has undergone significant changes since the end of apartheid. In addition to reentering world markets, significant restructuring took place, and also a significant reorientation in the relationship to government actors. While the South African mining industry was deeply entrenched with the apartheid government, new forms of interaction are currently emerging that present an interesting subject for study.

In terms of the governance context, South Africa is an emerging economy with a significantly developed industrial sector and also a relatively stable national-level government. At the same time, variation at the provincial level allows for an analysis of the role of government capacity, represented by the situation in two of the nine South African provinces chosen for the case study. Government capacity ranges from relatively well-developed institutions in some provinces to undercapacitated administrations in others. There is also a detectable gradient in terms of government capacity between the national as well as the local government level, with significantly weaker institutions in the latter case.

For this study the focus is placed on a close investigation of two mining areas located in two South African provinces: coal mining in the Mpumalanga province and platinum mining in the North-West province.

Coal mining in the Mpumalanga Highveld is a relatively stable and well-established mining sector (Coetzee, 2008). The water challenge here revolves around issues of water quality and security of water supply in neighboring communities. Those communities are put under increasingly severe pressure due to progressing climate change phenomena. The analysis for this region is focused on three mining companies, two of them subsidiaries of multinationals, one of them a smaller, emerging company.

Platinum mining in the Western Limb of the Bushveld complex in the North West Province represents a growing mining sector, with a high impact and high expectations in terms of its potential contribution to future economic growth (Coetzee, 2008). The water challenge is mostly related to water scarcity, access to water for local communities, and the need for regional infrastructure development, all of which are under increasing pressure due to progressing climate change. The companies considered in this case study area are three large mining firms and one smaller, emerging firm.

A qualitative and case-oriented research approach is used to explore the intricacies of business interactions with sustainable water management in these two areas. Involving a small number of cases, this method allows for an in-depth analysis of processes and social structures (George and Bennett, 2005), while deliberately limiting the potential for case-based generalizations (Ragin, 2000).

Data were collected during on-site studies in South Africa in 2008. The main sources of information were semistructured interviews with representatives from mining companies, government authorities, community stakeholders, and experts. Interview data were transcribed and analyzed according to the categories introduced in the previous sections.

Main Findings

The section that follows discusses the contributions of the case study's mining firms to sustainable water management and potentially to climate change adaptation. It also explores the firms' apparent motivations for the actions they have taken.

Case Study Area I: Coal Mining

In one example of company initiatives in the coal mining area, corporate actors developed and provided the technological solution to a pressing water pollution problem that had been caused by the firms' own operations The solutions however had much wider implications. In a collaborative industry initiative, firms in the area provided for the funding and realization of a state-of-the art treatment facility, a water reclamation plant to treat harmful acid mine drainage. The treated mine water is pure enough to be provided to the local municipality for drinking purposes, thus giving the municipality a relatively low-cost way of obtaining additional water to augment scarce municipal supplies. The plant represents a ten-year experimental project from a technical point of view, but also involved numerous negotiations with regulators, municipalities, and other affected stakeholders (Günther, Mey, and Van Niekerk, 2006). Only through leveraging corporate expertise and finances, and this web of negotiations, could the project have become possible.

This contribution was achieved through a collaborative effort of two of the largest mining companies in this area. The reclamation project is flanked by monitoring and planning activities, which take place at the individual firm level, or are undertaken as a joint effort with local partners, thus also addressing planning issues at the regional or catchment level. These collaborative initiatives have received positive assessments by regulators and are recommended as best practice to other operators. Positive results in capacity-building at the local municipality level were also achieved in this project. Through a trustful cooperation with the mining companies directed at a common goal, capacities for managing water-related infrastructure as well as regional water resources as a whole were built up with municipal managers as well as with regional water authorities (Kranz, 2010). It is important to stress, however, that such contributions are usually provided by only the larger, well-resourced firms, while emerging companies are struggling merely to comply with existing regulations, putting this type of project beyond their present capabilities.

In assessing the drivers for such positive company initiatives, regulation at the national level, even though not always enforced thoroughly, constitutes the most important driver. Regulatory pressure and the desire to avoid legal costs are in most cases augmented by concerns about company reputation and firms' strong desire to avoid reputational costs from negative social reception (Garner, 2008). This latter factor is especially important in leading to more proactive behavior among the major coal companies because the companies' reputations are increasingly tarnished by other impacts of the industry (e.g., coal combustion as a main contributor to climate change).

In situations like the one that led to the water purification plant, the response to immediate community needs is gaining in relevance and becoming an integral element of the incentive structure. This holds especially true in cases where the capacity of local government to provide for water-related services is hampered. However, the case study shows that some capacity needs to be present on the part of local government to enable successful cooperative approaches between municipalities and firms to occur. In this case, the local capacity did exist and led to the successful combination of mine water reclamation and municipal water supply.

A further driver is a variation of peer pressure. As mines are literally on each other's doorsteps in the Mpumalanga Highveld, environmental impacts are often not attributed to individual firms, but rather to the industry as a whole. For this reason, firms observe and assess each other's actions and the impacts of those actions quite critically, paying particular attention to the actions of the smaller mines. There is, however, also some competitive pressure among larger firms in the domain of technological developments. This observation hints at the important roles that internal factors play in shaping the specific behavior of firms. Organizational slack, the ability to manage economic resources, individual leadership, and the ability to collaborate with partners are also crucial in crafting successful contributions to addressing the water challenge (Kranz, 2010).

Case Study Area II: Platinum Mining

The contributions to addressing the water challenges in the second case, located in the platinum mining area, are also multifaceted. Internally, companies have put in place comprehensive on-site management systems that monitor water intake and usage throughout ore mining and ensuing processing steps. In addition, firms have engaged in developing technological innovations to increase on-site water use efficiency and recycling. Companies have also contributed to a better understanding of the water situation in this region by engaging in the monitoring of water flows. Initially, company activities were focusing only on production sites, but the scope of the activities has increasingly broadened to include surrounding areas as well.

The contribution with the greatest impact is clearly the stabilization of water supply for platinum mines and the resulting freeing up of potable water to boost availability for communities across the region. This greater availability of water is expected to alleviate water stress for local users as well as make a contribution from an environmental perspective. To support the development of these opportunities firms have engaged in a joint initiative.

In the first place, firms established an on-going infrastructure forum, a platform first to facilitate exchange and collaboration among corporate actors and then with local communities. The goal of the collaborations is to find joint solutions benefitting the firms and the local communities. In the context of this initiative, business is driving and contributing to the joint planning of infrastructure development and financing projects. The financing aspect is particularly relevant. Mines are not funding the construction of the infrastructure per se. Rather, capital is being leveraged from private lenders (i.e., banks).

In some projects, financing is being sought from banks based on contracts with the mines to off-take (industry-grade) water provided by a newly constructed pipeline. In addition, the mines will pay for the water at an appropriate price, thus covering upfront construction costs. Without the mines' "buy-in," the municipalities would not be able to get access to the funds necessary to build this infrastructure. While these projects have not yet been realized, a workable consensus has been found with the relevant local municipalities.

Parallel with their activities in the infrastructure forum, companies have been engaging in capacity-building measures for local municipality staff. These efforts are on-going and include addressing issues such as administrative and technical capacity. Similarly, multistakeholder platforms and partnership approaches have led to productive interactions between firms and local municipalities/communities that are likely to assist with addressing future challenges related to climate change impacts.

A decisive driving force in this case study is the dependence of companies on water as a key production factor. As a consequence they need to understand and secure how they manage and access water. These needs have led to the activities described above. Through their need for and use of water, firms are also linked to the needs and challenges of the surrounding municipalities. Firms are dependent on municipalities providing basic municipal services so they can maintain a reliable production environment. Municipalities on the other hand are faced with the challenge of enabling and supporting economic activities in the area, while at the same time fulfilling their constitutional mandate of ensuring sustainable livelihoods for their communities, especially under conditions of increasing climate change. Lack of, and even diminishing, capacity in varying degrees makes performing all these tasks a struggle in the districts studied.

The capacity-constrained situation of local municipalities and districts, combined with the need for securing water supply without compromising the municipal water situation, constitutes quite a strong driver for business and has resulted in the formation of the so-called Western Limb Producers'

Forum. This forum is designed to create a win-win situation for companies as well as communities. As they overcome previously insular approaches and establish continuous forms of interaction with municipalities targeted at capacity-building at the municipal level, corporate strategies are achieving greater possibilities of finding mutually beneficial solutions. At the same time, the collaborative structures also allow for the companies to share the risk in addressing the complex water situation in this region.

Regulation also constitutes a relevant driver for these company activities. The involvement of the national government in most instances has a mandatory character and has led to relatively good performance levels in terms of water-use efficiency, compliance, monitoring, and planning. However, under South African legislation provisions related to integrated development planning and water services development need to be implemented at the considerably weaker local-government level. This weakness at the local level poses a substantial challenge for integrating corporate water-related initiatives with planning efforts at the municipal-government level. While the regulatory framework is in place, the implementation is lagging behind. These factors are also prompting firms to adopt a proactive role with regard to water management in the region.

These external or institutional drivers resonate with a favorable internal disposition of firms toward addressing water challenges by relying on organizational slack and in some cases through internal leadership. In such cases well-developed sustainability management systems in the internationally oriented firms are drawn upon to support such leadership. Such company leadership is encouraged by a normative environment based on internationally established norms and historic legacies, as well as by community expectations. These community expectations feature quite prominently in supporting such company leadership, especially in combination with the leverage provided by traditional authorities and tribal communities present in that area.

In the next section, the case study results are assessed against the framework established in the previous sections. What are the activities and initiatives firms choose to engage in? In what way do they constitute a contribution to sustainable water management and thus adaptation to climate change? On the other hand, deficits are identified. Where does the framework not hold? What modifications are potentially necessary?

Overview of Findings and Refinement of the Framework

In the two case study areas investigated, firms engaged in quite similar *contributions* to sustainable water management, thus potentially reducing the risks and impacts of climate change in the local communities and improving climate change governance, albeit to different degrees and levels of involvement.

Firms in both cases were involved with planning and monitoring activities, with efforts focusing on internal water management questions in most cases and expanding beyond the company fence in some. The broader activities, which involve local and regional government departments or even catchment agencies, have the greatest potential in assisting with the detection of climate change impacts and thus provide for a profound basis for addressing these adequately. These activities and potential impacts relate to the information and preparedness requirements formulated above. It should be noted, however, that the firms' engagement in this domain is comparatively weak, compared to the often well-developed capacities of particularly mining firms in the hydrological and geological disciplines.

Technological innovations, leading to water efficiency improvements as well as reuse of waste water are often the strategy of choice. Reducing the use of water resources is a commendable adaptation strategy—one that benefits firms' bottom lines and also alleviates water stress in surrounding communities. Waste water treatment, on the other hand, constitutes an end-of-pipe technology that requires considerable investment and is in most cases highly energy-intensive. Therefore, the positive effects in terms of adaptation are potentially off-set by the creation of harmful climate-relevant emissions.

The infrastructure development and financing activities of the firms also have positive implications for the ability to manage scarce resources more effectively and to provide for an equitable allocation among all potential water users. Often municipalities, especially under conditions encountered in South Africa, lack the means and capacities to engage in infrastructure development and financing. Funding facilities as well as technological know-how provided by firms can thus help to alleviate these gaps. On the other hand, the long-term implications of infrastructure developments are often difficult to assess in terms of their comprehensive impacts on land and water management. In some cases they might hamper integrative water management and compromise sustainable development.

In the context of other beyond the fence activities, firms were also contributing to improving water and thus climate change governance. They did so through building up local administrative capacity to engage with corporate actors and to address water management issues more effectively. They also did so through awareness-raising and the initiation of multistakeholder dialogues for joint discussion of emerging issues. The latter activities might also have contributed to joint learning and thus the build-up of overall adaptive capacities for upcoming challenges.

The degree to which these activities will eventually lead to the improvement of the adaptive capacity in certain contexts, however, needs to be the subject of further investigations. Based on the cases investigated in the research, only a narrow assessment of the efficacy of firms' contributions to sustainability efforts can be made. If it is reasonable to assume that access to more complete information about water resources, strengthened institutional capacities, more careful infrastructure-related planning, and improved exchange with all relevant stakeholders will lead to an increase in adaptive capacities, then firms' contributions to sustainability efforts could be appreciable. It is worth noting, however, that most South African firms catered mainly to their own needs and put in only limited extra initiative to provide benefits beyond their own resilience vis-à-vis climate change impacts. The analysis of drivers and motivations leads to further clarifications in terms of the underlying dynamics.

Among the most decisive *drivers* of sustainability reform in the water management cases investigated are government actors, whether they are driving change through authoritarian approaches, such as regulation, monitoring and enforcement, or employing more facilitative approaches that seek to enhance cooperation with or among the firms.

Important firm-internal factors include accessibility of water resources, financial resources, and the overall disposition of the firm to address sustainability and water resource challenges. Detailed case analyses also display interactions among governmental drivers and other institutional drivers, such as social factors (NGOs, community groups, and traditional authorities), competitive drivers (market, prices), and normative drivers (international and national norms), which often exert an enhanced ameliorative effect.

A sequence of drivers could be identified in which certain drivers, such as government pressure and firm-internal considerations, lay the foundation for the introduction of other drivers. Next to strong government intervention, weaknesses of government actors (especially local ones) in delivering appropriate services or effectively addressing important issues can constitute either a barrier to firm contributions or a driver for proactive firm initiatives or both.

In the context of the South African mining industry, the most sweeping changes conducive to reducing climate change risks were made when firms cooperated with the government at the municipal level. This observation suggests that resource-based drivers (e.g., water scarcity) alone do not suffice as determinants of business contributions to climate change adaptation, but instead need to be complemented or channeled by the intervention of institutional drivers, most importantly government intervention.

It is useful to note that awareness of climate change as a threat and the necessity to adapt appeared to be only a secondary driver. Other drivers dominated under the conditions encountered in the case study areas. Water scarcity was however gaining in importance as a motivator and climate-change-related considerations are likely to become even more decisive in the future.

The framework used for this investigation was useful and called attention to specific factors important in the South African mining industry. It becomes clear that this industry is very much focused on technology- and infrastructure-oriented solutions and also displays a characteristic incentive structure, driven by resource considerations and government intervention. These aspects are discussed in the final section.

Conclusions and Policy Implications

Businesses have a crucial role to play when it comes to addressing climate change impacts. In addition to addressing and mitigating climate change emissions as part of their business strategies, companies can play a crucial role in climate change adaptation by increasing their own resilience to environmental issues (Linnenluecke and Griffiths, 2010). They can also contribute to overall preparedness for climate change by collaborating with their surrounding communities. The South African Mining industry offered insight into potential strategies for business-community initiatives, but it also highlighted some of the limitations of such initiatives. In the South African situations investigated, these limitations seem to be partly dependent on the motivational pattern of the firms and most decisively on the respective governance context. While the dependency and thus vulnerability of the firm itself to environmental change definitely plays a role in determining firm behavior, the role of government as a regulating and/or facilitating driver of change also has a strong effect on business behavior.

In conclusion, a combination of different drivers, organizational values, vulnerabilities to climate change, and resource dependencies of firms is helpful in eliciting a corporate contribution to addressing water resource challenges and thus climate change adaptation. The necessity for governments to play a strong role, to interact with a wide range of actors, and to be diligent in combining different policy tools remains a valid consideration. Those factors are especially important for achieving beneficial long-term results. However, the capacity necessary for governing bodies to fulfill their role may frequently need to be improved. Capacity-building measures are still warranted in areas of weak regulatory capacity and limited statehood and should be targeted not only to improve the financial viability of these measures, but also to improve the ability of the government to cooperate and engage with corporate actors and activate their internal motivations to take action.

Increasing firms' internal motivations to act might be achieved by highlighting shared vulnerabilities to climate change of companies and communities as well as shared responsibilities in adapting to these changes.

While the results of these case studies of the South African mining industry need to be validated by investigating dynamics in other industry sectors as well as other country contexts, they have clear implications for the debate on governance for sustainable development and the specific interactions of private actors, communities, and the state in the context of climate change adaptation in an emerging economy setting.

From a more global perspective, these findings shed a light on the on-going discourse on the role of business in a Green Economy. Businesses have a contribution to make to climate change adaptation and to overall sustainable development. How effective this contribution will be in delivering benefits for surrounding communities depends on the level of support firms receive and the incentives for action they face.

References

Addams, L., G. Boccaletti, M. Kerlin, and M. Stuchtey. 2009. Charting our water future—Economic frameworks to inform decision-making, *2030 Water Resources Group*: McKinsey and Company.

Ashton, P. J., and D. Hardwick. 2008. *Key challenges facing water resource management in South Africa.* Paper presented at the 2nd Biennial CSIR Conference, Pretoria.

Ashton, P. J., D. Love, H. Mahachi, and P. H. G. M. Dirks. 2001. An overview of the impact of mining and mineral processing operations on water resources and water quality in the Zambezi, Limpopo and Olifants Catchments in Southern Africa. Contract Report to the Mining, Minerals and Sustainable Development (Southern Africa) Project. Pretoria, South Africa; Harare, Zimbabwe: CSIR-Environmentek, Geology Department, University of Zimbabwe.

Bansal, P. 2005. Evolving sustainably: A longitudinal study of corporate sustainable development. *Strategic Management Journal* 26: 197–218.

Bates, B. C., Z. W. Kundzewicz, S. Wu, and J. P. Palutikof, eds. 2008. *Climate change and water: Technical paper of the intergovernmental panel on climate change.* Geneva: IPCC Secretariat.

Bullock, A., W. Cosgrove, W. Van der Hoek, and J. Winpenny. 2009. Getting out of the box—Linking water to decisions for sustainable development, *Water in a Changing World—World Water Development Report 2009.*

Coetzee, H. 2008. Personal communication, August 26. Pretoria.

Delmas, M., and M. W. Toffel. 2004. Stakeholders and environmental management practices: An institutional framework. *Business Strategy and the Environment* 13: 209–222.

Dietz, T., E. Ostrom, and P. C. Stern. 2003. The struggle to govern the commons. *Science* 302: 1907–1912.

Eisenhardt, K. M. 1989. Building theories from case study research. *Academy of Management Journal* 14(4): 532–550.

Fox, T., H. Ward, and B. Howard. 2002. Public sector roles in strengthening corporate social responsibility: A baseline study. Washington, D.C.: Corporate Social Responsibility Practice / Private Sector Advisory Services Department, The World Bank.

Garner, R. 2008. Personal communication, August 28. Johannesburg.

George, A. L., and A. Bennett. 2005. *Case studies and theory development in the social sciences.* Cambridge, MA: MIT Press.

Global Water Partnership—TAC. 2000. Integrated Water Resources Management, Vol. TEC Background Paper No. 4. Stockholm: Global Water Partnership.

———. 2004. Catalyzing change: A handbook for developing integrated water resources management and water-efficiency strategies. Stockholm: Global Water Partnership.

Gunderson, L. H., and C. S. Holling. 2001. *Panarchy: Understanding transformations in human natural systems.* Washington, D.C.: Island Press.

Günther, P., W. Mey, and A. Van Niekerk. 2006. A sustainable mine water treatment initiative to provide potable water for a South African City—A public private partnership, *Water in Mining Conference.* Brisbane, Queensland.

Hamann, R. 2010. Strategic change in organisations and governance systems in response to climate change as wicked problem: Framing a comparative research agenda. Cape Town: UCT Graduate School of Business.

Hart, S. L. 1995. A natural-resource-based view of the firm. *Academy of Management Review* 20(4): 986–1014.

Hoffman, A. J. 2000. *Competitive environmental strategy—A guide to the changing business landscape.* Washington, D.C.: Island Press.

Holling, C. S. 1978. *Adaptive environmental assessment and management.* Chichester: Wiley.

King, A. A., and M. J. Lenox. 2001. Does it really pay to be green? An empirical study of firm environmental and financial performance. *Journal of Industrial Ecology* 5(1): 105–116.

Kranz, N. 2010. *What does it take? Engaging business for addressing the water challenge in South Africa.* Berlin: Freie Universität.

Linnenluecke, M. K., and A. Griffiths. 2010. Beyond adaptation: Resilience for business in the light of climate change. *Business & Society* 49(3): 477–511.

Liu, J., T. Dietz, S. R. Carpenter, M. Alberti, C. Folke, E. Moran, A. N. Pell, P. Deadman, T. Kratz, J. Lubchenco, E. Ostrom, Z. Ouyang, W. Provencher, C. L. Redman, S. H. Schneider, and W. W. Taylor. 2007. Complexity of coupled human and natural systems. *Science* 317: 1513–1516.

Loorbach, D., J. C. Van Bakel, G. Whiteman, and J. Rotmans. 2009. Business strategies for transitions towards sustainable systems. *Business Strategy and the Environment* 19(2): 133–146.

Molden, D. 2006. Accounting for water use and productivity. In *SWIM Papers*, ed. IIMI. Colombo, Sri Lanka: International Irrigation Management Institute.

Mukheibir, P. 2008. Water resources management strategies for adaptation to climate-induced impacts in South Africa. *Water Resources Management* 22: 1259–1276.

O'Riordan, T., and J. Cameron. 1994. *Interpreting the precautionary principle.* London: Earthscan.

Ostrom, E. 1990. *Governing the commons: The evolution of institutions for collective action.* New York: Cambridge University Press.

Pahl-Wostl, C. 2008. Requirements for adaptive water management. In *Adaptive and Integrated Water Management*, ed. C. Pahl-Wostl, P. Kabat, and J. Möltgen. Berlin, Heidelberg, New York: Springer.

Pahl-Wostl, C., and J. Sendzimir. 2005. The relationship between IWRM and adaptive water management, *Newater Working Paper Series*, Vol. No. 3.

Pegram, G., and B. Schreiner. 2009. Financing water resource management—South African experience: EUWI, Global Water Partnership.

Pegram, G., S. Orr, and C. Williams. 2009. *Investigating shared risk in water: Corporate engagement with the public policy process.* Surrey: WWF-UK.

Pinkse, J., and A. Kolk. 2010. Challenges and trade-offs in corporate innovation for climate change. *Business Strategy and the Environment* 19(4): 261–272.

Ragin, C. 2000. *Fuzzy-set Social Science.* Chicago, London: University of Chicago Press.

Rogers, P., R. De Silva, and R. Bhatia. 2002. Water is an economic good: How to use prices to promote equity, efficiency, and sustainability. *Water Policy* 4: 1–17.

Schwartz, J. 2003. The impact of state capacity on enforcement of environmental policies. The case of china. *The Journal of Environment and Development* 12(1): 50–81.

Scott, W. R. 2001. *Institutions and organisations.* Thousand Oaks, London: Sage Publications.

Smith, C. N. 2008. Consumers as drivers of corporate social responsibility. In *The Oxford Handbook of Corporate Social Responsibility*, ed. A. Crane, A. McWilliams, D. Matten, J. Moon, and D. S. Siegel, 303–323. Oxford, NY: Oxford University Press.

Swanson, D. L. 2008. Top managers as drivers for corporate social responsibility. In *The Oxford Handbook of Corporate Social Responsibility*, ed. A. Crane, A. McWilliams, D. Matten, J. Moon, and D. S. Siegel, 303–323. Oxford, NY: Oxford University Press.

Turton, A. R., J. Hattingh, M. Claassen, D. J. Roux, and P. J. Ashton. 2007. Towards a model for ecosystem governance: An integrated water resource management example. In *Governance as a Trialogue—Government-Society-Science in Transition*, ed. A. R. Turton, H. J. Hattingh, G. A. Maree, D. J. Roux, M. Claassen, and W. F. Strydom. Berlin-Heidelberg: Springer.

Umweltbundesamt. 2001. *Nachhaltige Wasserwirtschaft in Deutschland—Analyse und Vorschläge für eine nachhaltige Entwicklung.* Berlin: Erich Schmidt.

Van Zeijl-Rozema, A., R. Cörvers, R. Kemp, and P. Martens. 2008. Governance for sustainable development: A framework. *Sustainable Development* 16: 410–421.

WBCSD. 2009. Water, energy and climate change. Geneva: World Business Council for Sustainable Development.

WBCSD, and IUCN. 2010. Water for business—Initiatives guiding sustainable water management in the private sector. Geneva: World Business Council for Sustainable Development.

Winn, M., M. Kirchgeorg, A. Griffiths, M. K. Linnenluecke, and E. Günther. 2010. Impacts from climate change on organizations: A conceptual foundation. *Business Strategy and the Environment* 20(3): 157–173.

Young, O. R. 2002. *The institutional dimension of environmental change: Fit, interplay and scale.* Cambridge, MA: MIT Press.

CHAPTER 7

Sustainability Reporting's Role in Managing Climate Change Risks and Opportunities

Ann Brockett and Zabihollah Rezaee

Introduction

In the post-2007–2009 global financial crisis era, financial information no longer satisfies the needs and demands of a broad range of stakeholders. Global companies are facing growing pressure—internally and externally—to manage their impacts on global sustainability issues effectively and to disclose relevant information about their business sustainability (BS) practices. Internally, corporate management is seeking information related to both financial and nonfinancial key performance indicators (KPIs) for all five areas of multiple bottom line performance: economic, governance, social, ethics, and environmental (EGSEE). Externally, nongovernmental organizations (NGOs), community associations, customers, suppliers, and employees want more information about corporations' long-term impacts on society and the environment.

Even traditional investors are recognizing the potential financial risks and bottom line impacts of all of the elements of EGSEE and seeking additional information to achieve better understandings of a corporation's risks and how those risks are being managed. For example, 83 percent of investors surveyed in 2010 by Institutional Shareholder Services, a proxy advisory firm, said they now believe environmental and social factors can have a significant impact on shareholder value over the long term as half of 2011 shareholder proposals centered on social and environmental issues (Ernst & Young, 2011a). Similarly, regulators such as the Securities and

Exchange Commission (SEC) are becoming more aware of the financial and reputational risks associated with global unsustainability. In early 2010, the SEC issued guidance reiterating the need and importance for companies to disclose their material risk associated with climate change (SEC, 2010).

Business sustainability and corporate social responsibility (CSR) have been extensively and inconclusively debated in the business literature and authoritative reports for many years. The question frequently being asked is: "Do corporations have a social responsibility to stakeholders other than shareholders, such as employees, customers, government, society, and environment?" Although the debate continues, corporations' primary goals are refocusing from short-term profit maximization to long-term increasing shareholder wealth, which includes properly managing all areas of EGSEE performance. These goals are being viewed through a new lens and are being redefined as creating stakeholder value, which encompasses fulfilling ethical, environmental, and social responsibilities.

While it is true that many corporations are seeking only to appease stakeholders in areas where significant pressure is being applied, corporations that truly seek to integrate consideration of appropriate stakeholder concerns into their strategy-making processes and to manage sustainability issues effectively find that doing so is an integral component of corporate governance, entirely consistent with increasing shareholder value. It is important to note that there need not be a conflict between corporate goals of achieving high levels of profitability and advancing social agendas. In fact, embedding sustainability into corporate business strategy can actually create a form of shared value. Businesses can create competitive advantage by combining environmental and social needs with improved business performance to spur innovation and boost bottom-line results.

Business Sustainability

Rezaee and Brockett (in press) define business sustainability (BS) as the process of moving away from a short-term focus on profitability to a business model based on generating enduring performance in all areas of EGSEE. This business model creates sustainable shareholder value while also managing the interests of a broader group of stakeholders, including customers, suppliers, employees, government, society, and environment.

The benefits of a BS program include addressing environmental matters, reducing waste, reducing risk, improving relations with society, and avoiding punitive regulatory actions. BS programs provide corporations with the necessary tools to make sustainable products, take proper actions to promote social good, and advance social goals above and beyond creating shareholder value or complying with applicable laws and regulations. BS programs should also promote a set of voluntary actions advancing the social good, which go beyond the company's obligation to its various stakeholders. BS activities should be transparent and measured the same way financial activities are measured and disclosed.

Five Areas of Sustainability Risk

The 2007–2009 financial crisis can be attributed to many factors, including inadequate risk assessment of business transactions. Risk management has become an integral component of managerial functions affecting every transaction and economic event.

It is important to note that, while no business can grow without taking proper risks in managing its operations—the types and levels of risk commonly known as appropriate risk appetites or risk tolerances—there is no clear guidance in determining those appropriate types and levels of risk.

To assist organizations in grappling with these questions, the Institute of Risk management (IRM) has published a useful consultation paper providing guidance for corporations and their board in determining the nature and extent of risks necessary in achieving strategic objectives (IRM, 2011).

The importance of systematic risk assessment can also be found in the Dodd-Frank Financial Reform Act of 2010, which requires organizations to focus on compliance risk that should be integrated into the risk assessment process. This inclusion in the Act underscores the importance of risk assessment and its integration into corporate governance best practices. The Committee of Sponsoring Organizations of the Treadway Commission (COSO) provides a framework for implementing Enterprise Risk Management (ERM) and suggests that implementing ERM protects and enhances shareholder value (COSO, 2004).

The International Organization for Standardization (ISO) published its new standard: ISO 31000: Risk Management—Principles and Guidelines in 2009, which provides principles and generic guidelines on risk management (ISO, 2009). These ISO 31000 guidelines assist organizations in the design, implementation, assessment, monitoring, and continuous improvement of risk management in operations, compliance, and reputation as explained in the next section.

Organizations need to identify risks and opportunities and use methods and processes that help them manage risks and take advantage of opportunities. The overriding objective of implementing ERM is to manage overall financial risk as well as the risks related to strategy, operations, reporting, and

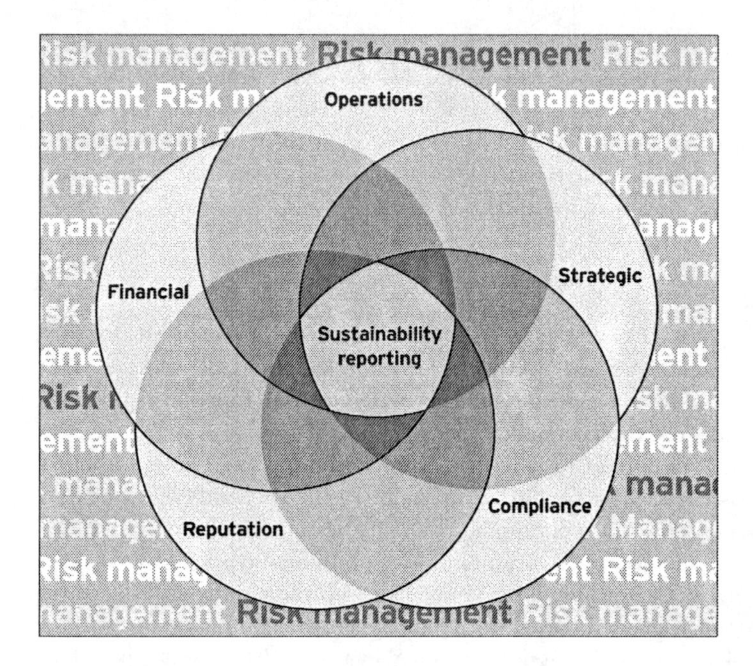

Figure 7.1 The five risks relevant to sustainability.

regulatory compliance. Managing the risks related to BS should also reduce organizational risk, while enabling the achievement of organizational goals and objectives including financial performance, market returns, return on investment, customer satisfaction, new products and services development, and stakeholder satisfaction. ERM enables organizations to (1) monitor risk and develop policies and procedures to address risk; (2) communicate risk assessment policies and procedures to managers throughout the organization; and (3) encourage managers to manage their risks relevant to their business strategies, operations, customer satisfaction, reputation, and stakeholder trust.

The five risks (strategic, operations, compliance, reputation, and finance) relevant to sustainability are depicted in Figure 7.1 and described in the sections that follow. Consideration of those five risks sets up the succeeding discussion of the increasingly important and effective role sustainability reporting plays in assisting businesses to improve their performance in all five EGSEE dimensions.

Strategic Risks

Businesses face a number of strategic risks associated with sustainability, from marketing position and changing consumer demand to strategic investments, stakeholder communications, and investor relations. Obviously, these risks will prompt management to focus on what could go wrong. But managers should also be focusing on where the opportunities lie. Businesses need to understand how they create competitive advantage by developing green products and services, increasing the efficiency of their operations, reducing energy and waste, and improving the communities in which they operate.

In a recent Ernst & Young survey, respondents indicated that the top three factors driving their climate change initiatives were (1) energy costs, (2) changes in customer demand, and (3) new revenue opportunities (Ernst & Young, 2010a). In other words, many executives know that a strong sustainability strategy aligned with their broader business strategy is a critical component of managing their business. For example, targeting energy costs can serve to not only reduce a corporation's environmental footprint, but also enable the savings that are achieved to be reinvested in other business initiatives within the organization. The same initiative can reduce the organization's risk exposure to regulatory requirements, such as greenhouse gas (GHG) emission reduction targets or other energy standards. Such an initiative will be particularly important (i.e., especially "strategic") for energy-intensive industries.

Changes in consumer demand as driving factors are self-evident. An ever growing number of informed consumers care about the environmental or social impact the products or services they purchase may have. New revenue opportunities could mean developing a line of green products or moving into new markets. All of these opportunities carry some form of strategic risk as well. Businesses that position themselves to manage both the downside risks and upside opportunities associated with sustainability have the strongest likelihood of being successful.

Operations Risks

One of the greatest challenges for companies in implementing their sustainability strategy revolves around collaboration and integration across operational business units and key functional areas. Such areas include IT, the supply chain, and production facilities. Sustainability-related issues include those associated with risks linked to the physical impacts arising from climate change. Climate change can lead to damage to infrastructure and assets, reduced asset life, and increased maintenance expenses, which in turn can interfere with operations, reduce performance in all EGSEE areas—especially company profitability—and increase insurance premiums. Another significant operational risk is the rising cost of energy, which is one of the largest operating expenses for many companies. A company's response to climate change and its carbon-reduction efforts—a risk area—is also an opportunity to reduce energy costs through implementation of efficiency measures.

Value chain risk associated with customers and suppliers is a growing operational risk area. Everyone is part of someone else's supply chain. Many corporations are working directly with their suppliers with the expectation that the suppliers will provide the customer with sustainability performance information, including their carbon footprint, water and waste information, and labor policies. Many corporations are now required to complete a life-cycle GHG assessment of their products and to provide this information to their customers. Many corporations are also being asked to disclose their plans for reductions in carbon content and costs. For all these reasons, corporations are now focused on their supply chains as both a risk area and as an opportunity to enhance operational efficiencies.

In a recently emerging development, the Global Reporting Initiative (GRI) has identified disclosure on sustainability issues across the supply chain as a revision priority to its Sustainability Reporting Guidelines. In October 2010 the GRI created an international multistakeholder working group, to make recommendations for improvements to its Sustainability Reporting Guidelines (GRI, 2010a). These recommendations are expected to address possible improvements of the quality of disclosure on performance of supply chains.

Compliance Risks

Many companies are facing new and expanding regulatory compliance risks resulting from an increasing number of international, national, and regional programs. In a recent study of 300 global executives from organizations with revenues of US$1 billion or more, 94 percent of respondents see national policies as important or very important in shaping their sustainability strategies and 81 percent recognize the importance of global or international policies in doing so (Ernst & Young, 2010a). In the past year alone, over 250 climate-change-related government actions were implemented globally—including state and provincial action across North America. In the United States, for instance, the Environmental Protection Agency declared carbon dioxide a danger to human health and announced that it would require reporting of GHG emissions starting in 2010 for heavy emitters—on a per-facility basis. As another example, the Federal Trade Commission has indicated its intentions to crack down on "greenwashing" in marketing claims (i.e., misleading consumers and stakeholders about environmental claims). These initiatives not only open up new regulatory compliance risks for companies, but also reputational ones, given that specific facilities will be placed under the microscope.

The key risk areas resulting directly or indirectly from regulatory measures are varied and can include health and safety, human rights and labor laws, anti-bribery, and environmental risks. Environmental risks can include direct impacts (e.g., emissions trading cost exposures) and indirect impacts (e.g., energy price increases and accompanying reporting and compliance costs). Audit and verification activities will also be required under certain programs, resulting in additional cost exposures. Companies in unregulated jurisdictions face additional risks around policy uncertainty.

In February 2010, the SEC published guidance regarding its disclosure requirements related to climate change risk. Issued in response to petitions from several institutional investors, the guidance does not amend any existing disclosure requirements or create any new ones. It does however signal companies to maintain a heightened awareness of climate change risk when preparing disclosures for SEC filings. The SEC's interpretive guidance highlights the following areas as examples of where climate change may trigger disclosure requirements (SEC, 2010):

- *Effects of legislation and regulation*—When assessing potential disclosure obligations, a company should consider whether the effect of certain existing laws and regulations regarding climate change is material. In certain circumstances, a company should also evaluate the potential effect of pending legislation and regulation.
- *Effects of international accords*—A company should consider, and disclose when material, the likely risks or effects on its business arising from international accords and treaties relating to climate change are likely to have.
- *Indirect consequences of regulation or business trends*—Legal, technological, political, and scientific developments regarding climate change may create new opportunities or risks for companies and may need to be disclosed.
- *Physical effects of climate change*—Companies also should evaluate, for disclosure purposes, the actual and potential material effects of environmental matters on their business.

Reputation Risks

One of the key drivers that many companies face is managing the expectations of a range of stakeholders—including investors, employees, customers, suppliers, local communities, and the media—and seeking to reduce risks to the companies' reputations and brands. Analysts are also starting to value companies based on their sustainability performance, creating new reputational risks. In the Ernst & Young study cited above, more than 40 percent of the respondents believe that equity analysts currently include climate-change-related factors in company valuations (Ernst & Young, 2010a).

Sustainability performance is also linked to customer satisfaction and loyalty, strong supplier relationships, and attracting and retaining top talent. Given the multitude of stakeholders exerting pressure on companies to manage climate change risks and to seize opportunities in this area, there has been a significant increase in external reporting. More than 1,200 companies worldwide now issue sustainability reports based on the GRI (GRI, 2010b).

Financial Risks

There is a growing awareness among institutional investors that sustainability issues can affect a reporting company's financial performance. Although not the only initiative of its kind, the United Nations Principles for Responsible Investment (UN PRI) is one of the largest: its 800 signatories manage more than US$22 trillion in capital and include large investment funds, such as BlackRock and TIAA-CREF. There has even been a proliferation of indices benchmarking the stocks of companies seen as sustainability leaders. These include the Dow Jones Sustainability Indexes (DJSI), the FTSE4 Good Index Series, and the NASDAQ OMX CRD Global Sustainability 50 Index.

Sustainability data are also available to institutional investors through commercial information services such as Bloomberg and Thomson Reuters, and to individual investors through websites such as Fidelity.com. Stakeholders can easily obtain a company's sustainability information and compare its reports with those of its competitors. More than 300,000 Bloomberg subscribers have access to comprehensive nonfinancial company information such as emissions data, energy consumption, human rights information, corporate policies, and board composition. Thomson Reuters gives more than 400,000 subscribers access to similar information at the touch of a button. Institutional investors and other users are reviewing sustainability data over time, comparing it across and within industry sectors, and sometimes ranking disclosure levels and reporting quality. For an organization's overall reputation, such scrutiny makes it important to maintain complete and accurate data on its sustainability practices.

Research also indicates that equity analysts increasingly consider sustainability practices when valuing and rating public companies. In mid-2010, a global survey of 300 executives in large companies showed that 43 percent believed equity analysts consider factors related to climate change when valuing a company (Ernst & Young, 2010a). More recently, a study by Ioannis Ioannou of the London Business School and George Serafeim at Harvard showed that equity analysts have begun giving higher ratings to companies with exemplary CSR practices (Ioannou and Serafeim, 2011). Ioannou and Serafeim surveyed more than 4,100 publicly traded companies over a 16-year period and found that since 1997, analysts have viewed CSR strategies as creating value and reducing uncertainty about future cash flows and profitability. As a consequence, in recent years the analysts have issued more favorable ratings to companies that have sustainability strategies in place.

Sustainability Reporting

Accounting for and reporting on BS is the process of identifying, classifying, measuring, recognizing, and reporting performance in all areas of EEGSE. Such reporting is referred to as "Corporate Social Responsibility or Sustainability Reporting" or "Triple-Bottom Line Reporting." Ernst & Young, one of the Big 4 accounting firms, defines sustainability reporting as "a process for publicly disclosing an organization's economic, environmental and social performance" and notes that the process has become more than a reporting exercise, but rather is looked at as a business opportunity that can identify additional revenue opportunities and reduce costs (Ernst & Young, 2011b, p. 2).

Sustainability reporting covers all areas of economic viability, ethical culture, corporate governance, social responsibility, and environmental awareness. Many companies follow the GRI Reporting Framework when deciding what to include in their sustainability reports. The framework starts with a series of principles that organizations can use to judge whether a particular piece of information merits inclusion in their sustainability reports (GRI, 2010b). The principles are as follows:

- *Materiality.* Information in the sustainability report should reflect the organization's most significant impacts on profit, people, and the planet in all areas of EGSEE performance described in this chapter. Material sustainability issues and EGSEE performance can be those that affect the organization's financial position in the short-term and long-term shareholder value creation. The guidelines emphasize that in assessing materiality of sustainability information attention should be given to the organization's overall mission, goals, and strategies as well as to stakeholder interests, broader social expectations, and environmental issues.
- *Stakeholder inclusiveness.* Sustainability reports should disclose and respond to stakeholders' expectations and interests. Stakeholders are typically any individuals or communities likely to be significantly affected by or have affects on what the organization does in achieving its goals and objectives.
- *Sustainability context.* The purpose of a sustainability report is to present an organization's commitments and actions in all areas of EGSEE

performance. The report can be prepared in an integrated format to serve a variety of purposes including financial, CSR, governance, and environmental matters.

- *Completeness.* Sustainability reports should reflect significant impacts of the business and enable stakeholders to assess its EGSEE performance in the reporting period.

The other global standard, AA1000AS (2008), was designed for use beyond the accounting profession. It was created by AccountAbility, a global nonprofit organization. AA1000 is a principles-based standard that, in addition to reported conventional financial information, addresses management and reporting systems and processes. Using both the International Standards on Assurance Engagements 3000 (ISAE 3000) and AA1000AS (2008) is considered to be a leading practice in sustainability reporting. Country-specific general assurance standards, such as the American Institute of CPAs' (2001) AT101 and the Canadian Institute of Chartered Accountants (2011) *Handbook* Section 5025, can also be used in nonfinancial reporting in their respective countries.

The International Integrated Reporting Committee (IIRC) was created in 2010 to assist in the development of an approach to integrate all of the elements of EGSEE performance and reporting by entities of all types and sizes (Integrated Reporting, 2011). This integrated report is intended to serve all stakeholders, be comprehensive and reflect financial and nonfinancial KPIs in all areas of EGSEE performance, be transparent and balanced, comply with all applicable laws, rules, regulations, and standards, and, very importantly, be based on an organization's sustainable strategies. The IIRC comprises representatives from Accounting for Sustainability (A4S); the International Federation of Accountants (IFAC); the Global Reporting Initiative (GFI); the major global accounting firms; prominent financial and accounting organizations; the United Nations (UN); the World Bank; the International Organization of Securities Commissions (IOSC); the International Monetary Fund (IMF); the International Accounting Standards Board (IASB); the Financial Accounting Standards Board (FASB); business, investing, and academic communities; and the Financial Stability Board as observer.

In the face of mounting pressure to be transparent, a growing number of organizations are choosing to report on sustainability and CSR as an integral component of their risk management procedures. Sustainability reports help readers understand how well the reporting organization adheres to the "triple bottom line" of social, environmental and economic (SEE) performance. Typically disclosed voluntarily, these reports also spotlight the sustainability-related risks and opportunities facing the reporting entity;

whether it's a public or private company, a government agency, an academic institution or a not-for-profit. More than 3,000 companies worldwide are currently preparing and disclosing sustainability reports according to an Ernst & Young (2011b) study. The companies include Microsoft, Cisco, Ford, Johnson & Johnson, Procter & Gamble, Best Buy, Chevron, ConocoPhillips, Xerox, HP, and Disney from the United States and Shell, BASF, ArcelorMittal, Novartis, Carrefour, Nokia, Siemens, HSBC, and Novo Nordisk from Europe.

An organization that reports on its sustainability practices is expected to show not only where it has succeeded, but also where it failed. This expectation creates an element of reputational risk in the short-term. However, over the long-term, the risk is outweighed by significant benefits: better measurement of the organization's "triple bottom line" performance, greater stakeholder trust, improved risk management, and increased operational efficiency—very likely facilitated to a considerable extent by the energies focused on remedying areas of reported weaknesses. Many of these benefits are derived from the internal processes and controls companies put in place to help them collect, store, and analyze financial and nonfinancial KIPs. Obtaining real-time, quality data on such issues as GHG emissions, water use, and supply chain activities can help companies enhance decision-making while reducing risk. Failure to report on sustainability, by contrast, can increase risk. Companies that do not release sustainability information may appear less transparent than competitors that do, coming across as laggards even if they are not. Furthermore, those that report incompletely, or with insufficient rigor, may find that if reporting becomes mandatory and standards are tightened, glaring discrepancies might appear between past reports and newer ones. All of these factors have created more openness and more transparent reporting.

Traditional financial statements, providing historical financial information concerning an entity's financial condition and results of operations as a proxy for future business performance, may be lacking in relevant information. Investors demand forward-looking financial and nonfinancial information on KPIs concerning the entity's governance, economic, ethical, social, and environmental activities.

Sustainability Reporting in Action

More than two-thirds of the Fortune Global 500 companies publish some form of sustainability report (Ernst & Young, 2010b). It is likely that a growing number of entities including corporations, not-for-profit organizations, and even private companies provide some sort of sustainability and accountability reports above and beyond conventional financial reports. Sustainability reporting focused initially on environmental issues in mining, utility, and energy industries but is now widely used in all industries including financial services, telecommunications, health care, logistics, and construction and even not-for-profit organizations. The study cited above also suggests that equity analysts increasingly consider sustainability practices such as climate change and CSR strategies when valuing and rating public companies.

As sustainability reporting gains widespread acceptance and is used by many high-profile companies in almost all industrial sectors, the reliability, relevance, and comparability of sustainability reports become increasingly important—particularly as rating agencies and financial analysts continue to utilize sustainability information in their ratings and valuations. Accurate and complete information is required because the information plays such an important role in determining the ratings and valuations, which eventually affect companies' cost of capital and stock values. Third-party audit and assurance reports on sustainability disclosures can be beneficial to companies in achieving their sustainability goals and to their investors, creditors, suppliers, customers, government, and society. Public accounting firms have traditionally provided audit reports on mandated financial statements and they are well qualified and prepared to provide assurance reports on all EGSEE dimensions of sustainability and accountability performance.

Although accounting and auditing standards for financial and audit reports reflecting economic performance are well established and practiced, sustainability reporting and assurance is still in its infancy, with many challenges and opportunities remaining for further refinements and widespread acceptance. The challenges include the establishment of standards. Globally accepted standards for measuring, recognizing, reporting, and auditing governance, ethics, social responsibility, and environmental activities and performance have yet to be established. Organizations may be reluctant to present unaudited KPIs on their ethics, social, governance, and environmental activities that may create expectations and further accountability for them to improve their performance in these areas. Another challenge is to find ways to disclose concise, accurate, reliable, complete, comparable, and standardized sustainability reports that are relevant and useful to all stakeholders.

Companies that are planning to initiate sustainability reporting activities in the near future should: (1) understand and assess the risks associated with climate change and global unsustainability; (2) consider the challenges, opportunities, risks, and rewards associated with disclosing sustainability reports; (3) make sure they are ready to address all relevant sustainability issues demanded by stakeholders; (4) design and implement appropriate accounting and internal control systems to measure, assess, recognize, disclose, and provide assurance on all EEGSE dimensions of sustainability performance; (5) obtain commitment from the board of directors and top-level management for the effective design and implementation of sustainability reporting; and (6) comply with all applicable sustainability laws, rules, regulations, and best practices.

Companies exploring the possible value of sustainability reporting would be well advised to address seven questions related to best practices, voluntariness, content, mechanisms, value relevance, and assurance of sustainability reports (Ernst & Young, 2010b). The questions are as follows:

Who issues sustainability reports (best practices)?
Why report on sustainability if you do not have to (voluntary)?
What information should a sustainability report contain (content)?
What governance systems and processes are needed to report on sustainability (mechanisms)?
What are the challenges and risks of sustainability reporting (assessment)?
Do sustainability reports have to be audited (assurance)? and
How can companies get the most value out of sustainability reporting (value-relevance)?

The Role of Assurance in Sustainability Reporting

There are two primary global assurance standards providing guidance for assurance on BS. ISAE 3000 is the benchmark that accountants most often use as a basis for assurance of sustainability reports. Although ISAE 3000 (2005) was developed by the International Auditing and Assurance Standards Board whose standards exist primarily to audit and review financial information, in 2004, the group also produced ISAE 3000 as a standard for nonfinancial assurance engagements. The other global standard, AA1000AS (2008), was designed for use beyond the accounting profession, and was created by AccountAbility (AA, 2008), as noted previously. Both ISAE 3000 and AA1000AS (2008) are considered to be a leading practice in sustainability assurance worldwide.

Country-specific general assurance standards, such as the American Institute of CPAs' (2001) AT101 and the Canadian Institute of Chartered Accountant (2011) *Handbook* Section CICA 5025, can also be used in nonfinancial reporting processes in their respective countries. Assurance standards on different dimensions of sustainability performance can vary in terms of vigorousness and general acceptability. For example, the Public Company Accounting Oversight Board (PCAOB) auditing standards in the United States or International Auditing and Assurance Standards (IAAS) governing reporting and assurance on economic activities presented in financial statements are well established and widely accepted and practiced. Assurance standards on dimensions of sustainability including governance, ethics, social, and environmental standards are yet to be fully developed and globally accepted.

Conclusion

Business sustainability (BS), which originally was viewed as a question of corporate governance, has now emerged as a central, multifaceted theme of the twenty-first century. It is now the responsibility of corporate boards and managers to focus on BS by creating enduring value for shareholders and managing the interests of other stakeholders, including creditors, employers, suppliers, government, and society at large. BS and corporate accountability is all about adding value in all areas of EGSEE matters and events. Many public companies now voluntarily manage measure, recognize, and disclose their commitments, events, and transactions relevant to EGSEE.

BS is a process of enabling organizations to design and implement strategies that contribute to enduring performance in all EGSEE areas. BS not only ensures long-term profitability and competitive advantage but also helps in maintaining the well-being of the society, the planet, and people. Where sustainability and the "triple bottom line" reporting are concerned, the bar is rising. To meet it, companies must begin moving away from reporting designed mainly to generate positive publicity, and toward more rigorous and externally validated communications that address real business issues and manage risk in the five key areas of strategy, operations, compliance, reputation, and financial management. Such a shift will require that management take a broader interest in the contents of sustainability reports and in how those contents are verified and communicated. This chapter provides organizations with ideas to rethink their overall business strategy by integrating sustainability development into their core values, actions, risk management, corporate governance, and corporate reporting. In the post-2007–2009 global financial crisis era, traditional financial information per se no longer satisfies the needs and demands for both financial and nonfinancial information on KPIs for EGSEE dimensions of BS.

Note: The opinions expressed in this article are those of the authors and do not reflect the opinions of Ernest & Young LLP.

References

AccountAbility (AA). 2008. *AA1000. Assurance Standard 2008.* AccountAbility Standard, 1–26. http://www.accountability.org (Accessed July 31, 2011).

Accounting for Sustainability. 2009. *Establishment of "An International Connected Reporting Committee."* London. www.accountingforsustainability.org (Accessed July 31, 2011).

American Certified Public Accountants (AICPA). 2001. *Attest Engagements AT 101.* AICPA, 1085–1115. http://www.aicpa.org (Accessed July 31, 2011).

Canadian Institute of Chartered Accountants. 2011. *CICA handbook: Assurance.* Toronto: Canadian Institute of Chartered Accountants.

Committee of Sponsoring Organizations of the Treadway Commission (COSO). 2004, September. *Enterprise Risk Management—Integrated Framework.* COSO, http://www.coso.org (Accessed July 31, 2011).

Ernst & Young LLP. 2010a. *Action amid uncertainty: The business response to climate change.* (EYG No. DK0054.) 2010, May 21. http://www.ey.com/Publication/vwLUAssets/Action_amid_uncertainty:_the_business_response_to_climate_change/$FILE/Action_amid_uncertainty.pdf (Accessed July 31, 2011).

———. 2010b. *Climate change and sustainability: Seven questions CEOs and boards should ask about 'triple bottom line' reporting.* 2010, October 27 (EYG No. FQ0015.) http://www.ey.com/Publication/vwLUAssets/Seven_things_CEOs_boards_should_ask_about_climate_reporting/$FILE/Seven_things_CEOs_boards_should_ask_about_climate_reporting.pdf (Accessed July 31, 2011).

———. 2011a. *Climate change and sustainability: Shareholders press boards on social and environmental risk: Is your company prepared?* (EYG No. FQ0021.) 2011, May 02. http://www.ey.com/Publication/vwLUAssets/CCaSS_social_environmental_risks/$FILE/CCaSS_social_environmental_risks.pdf (Accessed July 31, 2011).

———. 2011b. *TBL Technology considers sustainability reporting.* EYGM Limited. www.ey.com (Accessed July 31, 2011).

Global Reporting Initiative (GRI). 2010a. Supply chain disclosure. http://www.globalreporting.org/CurrentPriorities/SupplyChain/SupplyChainDisclosure/Supply+Chain+Disclosure.htm (Accessed July 31, 2011).

———. 2010b. *GRI Sustainability Reporting Guidelines,* Version 3.0 http://www.globalreporting.org/reporting framework (Accessed July 31, 2011).

Institute of Risk Management (IRM). 2011. *Risk appetites and risk tolerance.* IRM 1–44. http://www.theirm.org (Accessed July 31, 2011).

Integrated Reporting. 2011 International Integrated Reporting Committee (IIRC). http://www.integratedreporting.org (Accessed October 28, 2011).

International Organization for Standardization (ISO). 2009. *ISO 31000: Risk management–principles and guidelines, 2009. ISO.* www.iso.org (Accessed July 31, 2011).

Ioannou, I., and G. Serafeim. 2011. The consequences of mandatory corporate sustainability reporting, Research Working Paper No. 11-100, Harvard Business School. http://ssrn.com/abstract=1799589 (Accessed July 31, 2011).

Rezaee Z., and A. Brockett (in press). *Business sustainability and accountability: The emerging performance and reporting paradigm*. Hoboken, NJ.

Securities and Exchange Commission (SEC) (Monday, February 8, 2010). 17 CFR Parts 211, 231, and 241 Commission Guidance Regarding Disclosure Related to Climate Change; Final Rule. Federal Register, Vol. 75, No. 25. www.sec.gov (Accessed July 31, 2011).

The International Standard on Assurance Engagements 3000 (ISAE 3000) (January 1, 2005). *Assurance engagements other than audits or reviews of historical financial information. ISAE, 1–57*. http://www.accountability21.net (Accessed July 31, 2011).

CHAPTER 8

Climate Change Risk and Informal Recycling: An NGO and Private Sector Partnership in Bogotá

Candace A. Martinez

This chapter describes the crucial and often unrecognized role that developing world informal recyclers play in reducing the risks associated with climate change. "Project Pennsylvania," (PP) a pilot program launched by four multinational corporations and a nongovernmental organization in Bogotá, Colombia, serves as an illustrative example highlighting how innovative recycling partnerships can create value and provide environmental, economic, and social risk reduction as they contribute to efficient solid waste management systems in low-income countries.

By leveraging the expertise of informal recyclers, "pro-poor" (UN Habitat, 2010:198) civil- and private-sector collaborations reduce ecological risks by cutting greenhouse gas emissions in overstressed landfills. The collaborations diminish the social risks associated with globalization, such as stark income disparities between socioeconomic groups and the further disenfranchisement of marginalized populations. They also lessen business risk by supplying firms with steady supplies of less expensive recycled materials, by reducing both CO_2 impacts of industrial value chains and the flows of waste material to landfills or other disposal sites.

Inclusive models, such as PP, suggest that when governments allow informal recyclers to become partners in private enterprises' solid waste management initiatives, firms and governments in the developing world can achieve both sustainable *competitive* advantages and sustainable *development* advantages.

Introduction

The World Economic Forum recently cited ecological degradation, income inequality, and the marginalization of poor populations as the principal environmental, economic, and social challenges facing the majority of low-income countries in the twenty-first century's globalized landscape (World Economic Forum, 2007). Although these three parts of the sustainability trinity are intrinsically linked across the developing world, the consequences of failing to deal adequately with the first—climate change—are purportedly direst for Latin America. As Laura Tuck, the World Bank Sector Director for Sustainable Development in the Latin America and Caribbean region, has noted, the region produces approximately 6 percent of the world's global greenhouse emissions (Tuck, 2009). Yet, if worldwide emissions are not brought under control, it is forecast that Latin America will suffer a disproportionate number of negative impacts. According to a December 2009 report from the Economic Commission for Latin America and the Caribbean, if the planet were to continue at its current rate of producing greenhouse gas emissions, Latin America could see its future gross domestic product drastically reduced. Other potentially dismal outcomes for the region include melting glaciers in the Andes mountains devastating farmlands, rising sea levels causing mangrove forests to disappear, degraded lands claiming from 22–62 percent of Bolivia, Chile, Ecuador, Paraguay, and Peru, and the biodiversity of Colombia and Brazil becoming severely depleted (Estrada, 2009).

While international organizations like the United Nations and its Kyoto Protocol have defined the parameters of global warming and its potential negative effects on the planet, and while developed and developing national governments from around the world have pledged their support to help decrease their countries' contribution to the collective carbon footprint, the successful implementation of these grand plans ultimately depends on how efficiently things can get done on the ground, within countries and within cities. Supranational policy often ends up being played out in small, incremental initiatives that tend to be grass roots, political, and messy. This situation begs the question: How conducive is a country to enact environmental policies given the complex web of national and municipal regulatory and political regimes on the one hand, and the special interests of powerful groups on the other hand? In light of Latin America's particular vulnerability to the negative externalities associated with the climate change crisis, the programs that the region's countries establish to halt the rate of global warming take on a heightened significance, as do the sectors that are most affected by compliance.

In this context, the World Economic Forum Report has identified the improvement of solid waste management in developing countries as one of the most effective weapons with which to prevent an environmental Armageddon (UN Habitat, 2010). The UN-sponsored inaugural meeting of the Regional 3R Forum in Asia, held in Tokyo, Japan, in November 2009, echoed the importance of solid waste management in meeting global sustainability objectives and specifically targeted the critical role of recycling:

> . . . (At this international meeting) the need for reorienting production and consumption patterns at all levels towards sustainability have been highlighted, with (an) emphasis on waste management and giving the highest priority to waste prevention and minimization by encouraging the production of reusable consumer goods and biodegradable products and developing the infrastructure required to reduce, reuse, recycle (3Rs) and dispose in an environmentally sound manner . . .. (Tokyo 3R Statement, 2009)

Anna Tibaijuka, Under-Secretary General of the United Nations, concurs and adds that "spectacular results" are possible when developing country cities include informal sector recycling as drivers of their solid waste management systems:

> Many developing and transitional country cities have active informal sector recycling, reuse, and repair systems, which are achieving recycling rates comparable to those in the West, at no cost to the formal waste management sector. Not only does the informal recycling sector provide livelihoods to huge numbers of the urban poor, but they may save the city as much as 15 to 20 per cent of its waste management budget by reducing the amount of waste that would otherwise have to be collected and disposed of by the city. (UN Habitat, 2010: v)

The sustainability benefits of incorporating informal recyclers into a city's formal waste collection and disposal operations seem self-evident. What remains less clear, however, is the reluctance of many governments (municipal and state) in underdeveloped countries to draft and enact laws that would provide transparent and equitable regulatory frameworks for recycling activities. Such regulatory frameworks would allow informal recyclers to gain legal recognition (and potential remuneration) as contributing members of their cities' solid waste management efforts as well as contribute to cities' recycling objectives. A dearth of legislation that regulates, enforces, and protects recyclers and their right to work in the solid waste management sector currently defines the status quo in most developing world cities. In the majority of these cities, informal recycling workers (formerly referred to as scavengers or waste-pickers) are neither protected by law nor encouraged to contribute to the recovery of recyclable materials, an activity whose global impact has been estimated in the billions of US dollars (Medina, 2009).

Defining sustainable development as growth that fosters environmental, economic, and social benefits for present and future generations, it is difficult to imagine an economic activity that more successfully marries those three dimensions than do integrative solid waste management programs, that is, waste collection and disposal plans that include informal recycling. Nevertheless, cities and countries in the developing world have struggled to come to grips with this reality. Despite the potential of a legitimized (i.e., legalized) informal recycling program's ability to reduce costs and risks for stakeholders, the entrenched interests of powerful groups, a blurring of the lines between private profit and public policy, and the historic social marginalization of and discrimination against waste-pickers have contributed to many developing countries' failure to embrace and leverage the role that the informal recycling sector can play in efficient municipal solid waste management services (Dias, 2000; Guillermoprieto, 1990; Medina, 2005, 2007, 2008a, 2008b; Nas and Rivke, 2004; Schamber and Suarez, 2007).

Colombia and Bogotá are good examples of this tension. Located in the region of the world that stands to lose relatively the most if the natural environment's pollutants are not brought under control, they represent a strategic battleground where former waste-pickers (now recyclers) have long struggled for social inclusion legislation. This country-city dyad also exemplifies, however, the contradictory national and municipal public policies toward informal recyclers' legal status within the solid waste management sector. Colombia as a nation has recognized and legalized informal recycling (albeit in fits and starts), but enforcement mechanisms have been lax. As a result, the capital city lags in the adoption and implementation of the legally mandated measures intended to bring about full integration of informal recyclers and recycling associations into the mainstream of municipal sanitation.

PP represents the type of strategic alliance in the realm of the possible in those developing country cities where the spirit and the letter of the law allow inclusive partnerships that cross business and civil sectors. This chapter begins with a historical account of the legal trajectory of Colombia's informal recyclers to help readers gain an appreciation of the valuable contribution PP is making toward achieving Colombia's and Bogotá's recycling goals. This historical overview underlines the social significance of the movement from informal garbage dump scavengers to legalized entrepreneurial recyclers, a transition that underpins the regulatory context of the partnership. The chapter then introduces the project's operational design and its collaborating actors (the Bogotá Association of Recyclers and the four participating companies). A discussion of how the cooperative efforts of the alliance partners contribute to reducing sustainability risks is then provided. The chapter concludes with an assessment of what is required to scale integrative models, like PP, in other settings.

The Regulatory Framework for Recycling in Colombia

Projects like *Proyecto Pensilvania* are not born in a vacuum. Municipal and national governments proscribe the legal boundaries for all organizational and individual behavior within the confines of a country and a metropolitan area, and they define the competitive rules of the game for business activities within these constraints. Further, a solid regulatory foundation bestows legal, social, and political legitimacy. In Colombia and elsewhere in the developing world, scavengers (waste-pickers) have historically occupied the bottom rung of the socioeconomic ladder. Poor, marginalized, and uneducated, they have consistently been discriminated against by mainstream society and in Colombia have even been victims of criminal violence and of *limpieza social* or social cleansing campaigns. The Bogotá Association of Recyclers (or ARB as it is known by its Spanish initials) has struggled for more than two decades to gain recognition and legitimacy for its members, to be allowed to compete for municipal recycling contracts, and to form legal business ventures with the private sector. Several significant pieces of national legislation have contributed to defining the legal space that today governs the informal recycling associations in Colombia and delineates the rules and regulations for their operation and for their formation of private solid waste management partnerships. While these legal measures governing the ARB and other regional associations throughout Colombia seem to be intended to favor the formalization of informal recycling, they have yielded an ambiguous and confusing regulatory space with counterproductive results on more than one occasion.

In the 1990s a trend was becoming prevalent in Latin America for solid waste management to change hands from the public sector to the private sector (Johannessen and Boyer, 1999). This privatization became formal in Bogotá in 1994; Law 142 privatized public sanitation in Bogotá, from the collection, transportation, and disposal of solid waste to the cleaning and sweeping of city streets, to the oversight of the cleanliness of public squares. In 1999–2000, Law 511 and Decree 2395 were passed by the Colombian Senate, mandating that informal recycling, or "scavenging" as it was still known, be legalized. That is, Colombian law recognized the human right of scavengers to earn a living wage by going through the discarded waste in garbage dumps or on streets and selling the recycled goods for profit. In 2002 then President Pastrana issued a decree, Decree 1713, that subsumed and regulated all previous national norms relating to informal recycling and conferred legal status once again on informal recyclers. The decree went further by recognizing the value that informal recycling actors contributed to the environment. What this decree did not include were specific guidelines on the incorporation of recycling associations composed of formerly informal recyclers into cities' Master Plans on Solid Waste Management (Medina, 2007; Parra, 2007; Samson, 2009a). It did include, however, a controversial article later appealed by the ARB and overturned by Colombia's Supreme Court stipulating that all trash left on the streets became the private property of the consortium owning the recycling concession in that locale, insinuating that if an informal recycler or member of the ARB were to take any of it, s/he could be accused of stealing (Parra, 2007:77; Samson, 2009b:75).

Despite the national level pro informal-recycling laws and the fact that Bogotá's recycling associations are responsible for recycling approximately 600 tons/day of the city's waste (Parra, 2007:65), the ARB was still excluded from participating in citywide competitions for municipal waste management contracts. In 2003 when Bogotá sponsored its first "open" competition for the city recycling concession subsequent to Decree 1713, the ARB planned to submit a bid to cover part of Bogotá's recycling needs. To its dismay, the association learned that it was banned from participating in the public tender for various reasons. For large urban areas like Bogotá, only incorporated entities could provide residential waste collection and disposal; associations like ARB were restricted to rural areas or large towns. Further, only those legally formed companies with verification of US$5 million financial backing and with at least three years' experience in large solid waste management projects would be allowed to offer their tender. This ruling virtually eliminated the ARB's chances to compete for municipal solid waste management projects. In effect, the law was a Catch-22: the ARB members did not have any experience in organizing and operating a large-scale recycling project at the city level because the municipal government had never allowed them to organize and operate a large-scale recycling project at the city level. The ARB was particularly frustrated by the city government's attitude toward informal recyclers and recycling associations because, when called upon to help out the city when the then state-run street cleaners and trash collectors went on strike in 1994, the ARB answered the government's call to step in and clean Bogotá's streets and to pick up and transport its garbage to the city dump. It collected about 700 tons of waste per day during the strike, thus demonstrating to the government almost ten years before the 2003 competition that the ARB was, indeed, capable of managing important quantities of solid waste.

The ARB and its pro-bono legal team appealed what they perceived as discriminatory exclusion not only in the application of the 2002 law as applied to municipal recycling contracts but also in the application of the original 1994 law that privatized solid waste management. Citing their basic human right to be able to compete as any other legally organized entity, the ARB team brought their case to the *Corte Constitucional de Colombia,* Colombia's Supreme Court, and eventually won. The *Corte Constitucional* ruled, however, that the ARB would have to wait until the next public tender for the city's recycling contracts, set for 2010–2011. Another ruling by Colombia's Supreme Court affecting the informal recyclers of Cali was also decided in that city's Association of Recyclers' favor, but implementation has been slow and inconsistent (Samson, 2009b; Ruiz, 2011, Martinez, 2011).

Project Pennsylvania

It is against this backdrop of ambiguous public policy that the innovative, private- and civil-sector joint project *Proyecto Pensilvania* (Project Pennsylvania or PP) was launched in March 2010. For the first time ever in Bogotá, four multinational companies and the Bogotá Association of Recyclers formed a partnership to establish and operate a recycling center on the outskirts of the 7 million inhabitant city. In broad strokes, PP organizes the recycling of discarded materials and sells them directly back to industry, thus reducing costs through the elimination of recycling intermediaries. The motivating idea behind the project is to leverage the recycling experience and know-how of the ARB with the business expertise of the four partner companies and to achieve mutual benefits by doing so. The synergistic partnership showcases the efficiency gains of recycling, demonstrating that recycling associations, working together with industry, can provide a practical, affordable, and sustainable contribution to the city's solid waste management strategy.

PP's operational model is simple. About 50 recyclers, members of the ARB, form the core of PP's employee base. They collect six types of materials from their habitual recycling routes in the city's residential and business neighborhoods (paper, aluminum, glass, cardboard, Tetra Paks, and other cartons) and store their goods in their own *bodegas* or small warehouses. Three times a week an ARB truck stops at the 50-odd different recycling *bodegas,* picks up and weighs the bags of recycled goods, records their weight, type, and provenance and transports them to the *Centro de Acopio Pensilvania,* the Pennsylvania Warehouse and Transfer Station, located about

15 miles southwest of Bogotá's city center. If the individual recyclers or their associations do not have small warehouses where they can temporarily store their weekly cache of recyclables, then the ARB truck makes arrangements for pick up at designated times and locations at no cost to the recycler. By transporting their goods to a large warehouse, the ARB replaces the traditional middleman in the informal recycling value chain, the intermediary who pays recyclers pennies on the dollar for their individual bags of recycled goods that are later consolidated into large quantities and sold for a profit to industry, thus exploiting economies of scale.

Once the materials arrive at the Pennsylvania warehouse, approximately 14 workers, also employed by ARB, prepare the goods for industry purchase. They verify their weight, sort them into groups (glass, cardboard, et cetera), then wash, clean, and compress the recycled goods into compact blocks resembling bales of hay. The companies that have signed contracts with PP to purchase determined tonnages of the recycled materials send their trucks daily to pick up their agreed-upon type of recyclable. The seven partner companies with contracts to buy from the center (Grupo Familia Sancela, Fibras Nacionales, Diaco, Peldar, Empacor, Col Recicladora, and Cartonal) pay according to total weight upon pick up. Tetra-Pak also buys back tetra-pack containers, making it a client firm as well as part-owner of the operation. For their part, all ARB employees of the PP are paid in cash weekly on Saturday mornings, an unusual system for workers accustomed to receiving payment by intermediaries almost immediately after collecting their day's worth of goods. In addition to the free transport of their material to the *centro de acopio* and a living wage, the PP-employed recyclers also receive full benefits (including health insurance), uniforms, and "dignified treatment and professional status" (as the manager of the project phrased it to the author). As of early 2011, they are also given a weekly bag of groceries donated by a Catholic Church-sponsored food bank that serves as a link between recyclers' needs and donor firms.

The turnaround time at the Pennsylvania Warehouse and Transfer Station is 24 hours; it is filled and emptied daily.

The Bogotá Association of Recyclers (ARB)

The Bogotá waste-pickers' transformation from independent, self-employed scavengers who eked out a livelihood in the cities' landfills and streets into recyclers who belong to a formal, legal association, who lobby aggressively and tirelessly for government legislation and social inclusion policies, and who can participate as equal partners in solid waste management projects grew out of a confluence of events starting almost three decades ago. These events mirrored other cities' initiatives. Medellín and Cali, for example, both witnessed the formalization of their informal recycling workers in the mid-1980s as the Colombian government began to shut down open garbage dumps and construct "sanitary" landfills (banning access to scavengers and using more modern techniques for waste disposal and treatment). In Medellín the government's closure of one of the city's largest dumpsites eliminated a source of income for approximately six hundred families who had worked there. Thus was born the first scavenger—now recycler—cooperative of the country, made up of those ousted waste-pickers who were willing to establish and operate within the confines of an organized movement. Today Medillín enjoys one of the most expansive, organized, and government-supported informal recycling movements in the country.

Bogotá's informal recycling movement followed a similar path toward legitimacy. The closing of one of Bogotá's largest landfills in 1990 was the catalyst for the establishment of four recycling cooperatives and for their joining forces with *Fundación Social*, a Catholic NGO that provided financial, educational, and moral support to help scavengers organize. *Fundación Social* was an active proponent of the informal recycling movement, providing loans, training programs, and subsidies until it closed its doors in 1996 due to financial problems (Medina, 2007:156–159). Together the newly formed cooperatives and *Fundación Social* protested the landfill's closing. Although they did not succeed in this aim, they were successful in establishing the ARB as a cooperative in 1990 and subsequently in giving the city's informal recyclers a strategic direction with specific objectives: improving their earnings, eliminating predatory intermediaries, removing the social stigma associated with their vocation, and gaining recognition as legitimate contributors to the city's solid waste management solutions. The founding of Colombia's National Association of Recyclers (ANR) soon followed with a current national membership upward of ten thousand. Many of Colombia's city-level associations (such as the ARB) have signed formal contracts with the ANR so that their activities can be coordinated at the national level (Medina, 2007).

The term "cooperative" has become a misnomer. The government of former President Uribe passed a law in 2005 that banned the formation of cooperatives, claiming that private firms were taking advantage of this form of governance structure to avoid taxes and to more easily fire employees (Ruiz, 2011). Consequently the ARB is now a type of über-association, not a cooperative, although it is frequently still referred to as a cooperative or as a "co-op association." Having the same legal structure as a privately held firm, the ARB's membership is currently comprised of 24 associations whose individual members represent about 10–15 percent of Bogotá's

18,000–20,000 informal recyclers (Parra, 2007). The organizational chart reveals an elected president and a top management team that conducts training sessions, oversees projects, monitors recycler compliance on designated recycling routes, builds membership, and promotes and participates in national and international events in Latin America and in other regions of the developing world. The ARB's responsibilities for the Pennsylvania Project include hiring, overseeing and paying personnel, maintaining close communication with the on-site director of the warehouse to solve problems as they occur, acting as a liaison between the warehouse recycling operations and the clients' needs, and guaranteeing that all contractual agreements are met. Driven by industry requests, the ARB has also negotiated with the government to allow its members to take training courses in handling industrial recycling and hazardous waste, knowledge they can apply to PP and elsewhere.

While PP represents one of the ARB's achievements with industry, ARB is also involved in an ambitious, citywide recycling project with the Bogotá municipal government. Located at the city's largest transfer station called *La Alqueria*, members of the ARB recycle about 15 tons/day of the 600–700 tons of trash that are recycled daily in Bogotá, according to an estimate that Silvio Ruiz, former president of ARB, provided to the author (Ruiz, 2011). The ARB aspires to have all of the large-scale recycling of residential waste in Bogotá follow the model currently in use at the *Alqueria*.

The Multinationals

The launching of PP as a pilot program in partnership with the ARB was the brainchild of four multinational corporations: Carrefour, Grupo Familia, Natura Cosméticos, and Tetra Pak. All are members of CEMPRE, a socially oriented business organization founded in Brazil in 1991 and established in Colombia in 2008. Its vision focuses on raising public awareness of the benefits of recycling, helping former waste-pickers and informal recyclers become formalized into society, and fostering projects that require partnerships with informal recycling associations (CEMPRE, 2010). What these four firms have in common is a deep belief in their firms' corporate social responsibility mission, values that are upheld with financial and human capital.

The stated corporate objectives of the PP confirm this philosophy. The companies agreed that this project should reflect an efficient recycling model created by recyclers themselves with firm-level support in the background. The firms strived to build a recycling infrastructure that removed the role of intermediaries to the direct benefit of the recyclers. The corporate objectives also included introducing training programs for interested recyclers so that they could become certified in functional areas such as accounting and handling hazardous wastes. Each company gave financial support to establish the project and to organize the logistics. An appropriate space was found and rented; a truck was bought. For the start-up budget of US$90,000 ($180 million Colombian pesos), Natura Cosméticos contributed a little over half. The remaining corporate partners donated the rest. Originally uncertain of the project's ultimate realization of its goals, the firms initially committed for one year, but in view of its success—almost all goods are sold and the project is now self-supporting—neither the ARB nor the corporate partners foresee an expiration date, as long as the municipal government of Bogotá continues to allow them to operate in this competitive space. Indeed, the partners hope that this type of project will be duplicated by other firms elsewhere in Bogotá and in other cities in Colombia.

The four companies that comprise the *Proyecto Pensilvania* partnership are Carrefour, Grupo Familia, Natura Cosmeticos, and Tetra Pak.

Carrefour: A French international hypermarket chain with over 1,400 stores worldwide, Carrefour is one of the largest discount retail operations in the world. It is second in industry revenues only to Wal-Mart. With a presence in four Latin American countries, the company has been involved in efforts to promote sustainability for more than twenty years, focusing on two main areas: integrating sustainable development into its business activities and promoting sustainable development to its customers. Its programs in Bogotá have been both creative and successful. Among them was a store-sponsored recycling campaign in which ARB representatives worked at store sites to encourage shoppers to deposit their recyclable used products. Incentives were awarded to the consumers who deposited the most recyclable goods.

Grupo Familia: Grupo Familia is another example of a firm that has institutionalized its approach to sustainability and its deeply held social responsibility principles. A Colombian joint venture (50 percent Colombian ownership; 50 percent Swedish), Familia has a presence in over 20 countries around the world. It manufactures disposable personal hygiene products and describes itself as a "leader in the design, innovation, production, and distribution of personal care products." One of the key areas of concern in its social mission in Colombia is to dignify the life of informal recyclers, assign due value to their work, and to help achieve their social inclusion. The company dovetails this social mission with its business objective of securing recycled materials to use in its production. With a deficit of paper for their manufacturing plant in Brazil (where all their products are made), they currently have to import recycled paper for reuse in production. As one of PP's industrial customers, Familia purchases used paper.

Natura Cosméticos: Natura Cosméticos, arguably one of the most environmentally aware corporations in the world, has been an industry leader in respecting and in propagating respect for the natural environment since its founding in Brazil in 1969. In the cosmetics and personal care direct sales sector, Natura is considered a sustainability pioneer, from its upstream innovative sourcing of raw materials in the Amazon to its downstream inventive reuse and recycling of the plastic bags its associates (sales representatives) use to deliver customers' orders. According to one of the founding owners, Natura "has worked hard to integrate sustainability into its core strategy" (Ethical Corporation, 2008). Its mission in Bogotá, where it opened a facility in 2008, underscores the firm's steady and serious focus on its responsibility to society, concentrating specifically on the plight of one of the city's most marginalized populations, the informal recyclers. As well as its leading role in spearheading PP, Natura has also initiated several in-house programs that advance the work and improve the lives of Bogotá's informal recyclers.

Tetra Pak: Tetra Pak, a privately held company founded in Sweden in 1950 and headquartered in Switzerland, supplies hundreds of types of "bric" carton packaging to its industry customers. Operating in more than 170 markets around the world, 75 percent of the company's paper products are made from reused, recycled paper. Recently Tetra Pak has begun manufacturing furniture and other items made from recycled paper. Best known for its milk and juice cartons, Tetra Pak also has a deserved reputation for its sustainability projects and recycling practices. For Sustainable Development Week 2010 Tetra Pak partnered with Carrefour in France to promote responsible consumption by announcing the arrival of a certified eco-friendly carton package that will help protect biodiversity and foster responsible forest management. It has also partnered with the World Wildlife Federation, with the United Nations, and with the Prince of Wales International Business Leaders Forum to promote several of its international initiatives. In Colombia, its environmental sustainability focus is evident. Of the 14,000 tons of Tetra Pak that are recycled per year, 2,000–3,000 tons are recycled in Colombia. PP represents one of this multinational corporation's first programs in Colombia aimed at helping informal recyclers. Along with Familia, it, too, purchases recycled goods from PP, in this case, used Tetra Pak containers.

Project Pennsylvania's Contribution to Risk Reduction

PP has clear societal implications. The fact that four multinationals came together with the ARB to establish the PP partnership implies that they hold the deep conviction that socially desirable solid waste management solutions in developing countries need to create income opportunities for society's poorest subgroups. Concomitantly the project helps the four participating firms meet part of their social responsibility objectives, and, more important, it directly improves the lives of the recyclers that it employs. And, arguably it sets the bar higher for future private, civil, and public joint programs. PP also decreases some of the countless risks inherent in this informal profession—such as unhygienic conditions, risk of accidents, exploitation by middlemen, below minimum wage pay, uncertain working conditions, improper handling of hazardous waste, and possible exposure to toxic substances—and provides a clean and safe work environment, legal employment, and a steady livelihood with benefits (Martinez, 2010). For many who work at PP, it is the first time they have ever been paid a lump sum for a week's work and been faced with the challenge and opportunity of learning how to handle personal finances with a predictable income stream. PP also provides a logistical infrastructure that supports workers' day-to-day recycling collection, transportation, and disposal by incorporating them into a formal economic activity that is valued and recognized by law. PP further legitimizes the claim made by informal recyclers and recyclers' associations that their vocation is part of the solution in making the world more sustainable, not part of the problem. The project represents an important preliminary step toward the gradual recognition and eventual acceptance by society at large of ARB members as well as of nonmember, independent informal recyclers as crucial contributors in addressing one of the modern world's most pressing problems: climate change. In addition to the societal benefits that accrue to the informal ARB recyclers who participate in the program, PP also reduces business and environmental risks.

Business-Related Risks

The main business risk reductions that PP provides its industrial client firms are (a) a guaranteed steady flow of recycled materials; (b) lower prices for these recycled goods since there is no middleman at PP; and (c) indirect cost savings from using recycled goods versus virgin (primary) raw resources. For the country as a whole, PP highlights the benefits that an industrial ecosystem can provide. By recycling discarded material that is later transformed and reused in the manufacturing of new products, the process avoids waste, use of virgin materials, or both.

Like all client firms that purchase recyclables from PP, the recycled materials for reuse in manufacturing processes that Familia and Tetra Pak buy are secondary inputs that would otherwise have to be purchased on the open market at a steeper price. The firms also have a priori knowledge and informed expectations of the amount of paper or Tetra Pak containers that they can expect from the consistent recycled supply that Pennsylvania handles over time. These two operations management-type risks (quantity and quality) are not trivial and the cost and CO_2 emissions reductions are significant. It is estimated that using virgin raw materials in manufacturing requires 50 percent more energy usage relative to using recycled materials (Tellus Institute, 2008).

The PP also contributes to the other purchasing firms' environmental sustainability targets and their ability to adopt an industrial ecosystem model. An increasing number of environmentally conscious firms that seek to incorporate sustainability measures into their production facilities are adopting this approach. The industrial ecosystem concept refers to a closed value chain loop; the waste produced in the manufacturing process is captured for reuse and recycled into the production of new products (Medina, 2007:96). Instead of recycling in-house, a process most firms are not equipped to implement, the PP does the recycling for its business-to-business customers, thus allowing them to close their value chains and create ecosystems. While reusing 100 percent of materials may be an unattainable goal, more realistic objectives can, nevertheless, translate into cost savings.

If taken as a separate organizational entity that plays an integrative recycling role in solid waste management systems, the PP itself can be viewed as a kind of umbrella industrial ecosystem it helps to create. Although not a manufacturing facility that reuses its own recycled waste, it functions as an input-receiving site where raw recycled goods are transformed into usable, clean, classified tons of recycled materials ready for industrial producers to reuse as inputs in their own transformation processes. In this light, PP fulfills a significant role in creating and overseeing an integrated, closed-loop ecosystem.

Environmental Risks

The environmental risks that PP tackles are twofold and related. First, it contributes to Colombia's overall sustainability goals and its efficiency targets for improving nationwide recycling rates. While not as low as other Latin American countries' recycling rates (Colombia's is about 20 percent compared to Mexico's 15 percent and Chile's 14 percent), they fall well below developed country standards, such as the U.S. rate of 34 percent and Denmark's rate that approaches 70 percent (Godoy, 2010; EPA, 2009; UN Habitat, 2010). Additionally, as one of the 192 nation-states that signed and ratified the Kyoto Protocol's UN Framework Convention on Climate

Change (UNFCC, 2011), Colombia has pledged its commitment to combat global warming through limiting or reducing greenhouse gas emissions. The efficient collecting, separating, transporting, cleaning, packing, and selling of recycled materials for industrial use is part and parcel of every city's successful environmental sustainability goals. The principal components of an integrative solid waste management program—reducing, reusing, and recycling—all contribute to reducing the risks associated with climate change (Gutberlet, 2000). In Latin America and the rest of the developing world, however, recycling has not historically been a widespread practice, despite the developing world's rising use of disposable consumer goods that use more wrapping and packaging, and thus create more waste (Medina, 2007). The result is that more trash is deposited in landfills, generating higher levels of methane, a greenhouse gas that has 21 times the climate change impact of carbon dioxide, exacerbating the environmental damage from toxic leachates seeping into and contaminating water supplies, and putting a strain on already limited space in landfills.

The second environmental risk that PP addresses relates directly to this last concern. Landfills in the developed world tend to utilize modern, state-of-the-art techniques that comply with strict government regulation and that have been established to mitigate the dangers associated with garbage dumps (EPA, 2011). In contrast, landfills in the developing world are likely to have little to no government regulation and are often "open" (allowing entry by scavengers). Spontaneous methane-induced fires and myriad safety hazards are the norm (Cointreau, 2006). Although the trend toward modernization in the developing countries is to replace open garbage dumps with closed, sanitary landfills, a lack of government oversight is still a challenge (Project Design Document Form, 2006:10). How much waste is dumped, what kind of waste, how it is treated to avoid toxic pollutants and leachates from contaminating the water and air, what tonnage limits the space will allow—these remain concerns in so-called sanitary landfills. A growing population that increasingly mimics the developed world's production and consumption habits exacerbates these problems.

Bogotá's main landfill, Doña Juana, drives home this point. Privately owned and operated, Doña Juana opened in 1988 and each year is the destination of approximately 2.5 million tons of household waste or about 6,000–8,000 tons of solid waste a day (PDDF, 2009). The waste is buried in cells and the leachate (the toxic-filled liquid that is spontaneously created from decomposing waste), often untreated, is later discharged into a nearby river, not an atypical solution in developing countries (Hendron, 2006). Prior to the mid-1990s, the guiding municipal logic rationalized that every kilo of trash that was deposited in its final destination (Doña Juana) translated into one less kilo of trash on the streets (Parra, 2007:75). A systematic recycling program was not a top priority. A tragic landslide at Doña Juana in 1997 changed this reasoning. Almost 2 million tons of solid waste flowed to the river. While no injuries were reported, the environmental catastrophe (extreme pollution of the river, foul odors, illnesses, and epidemics) was massive enough to spur politicians into reconceptualizing their solid waste management policy for the city. They realized that landfills have a limited capacity and that alternative solutions to waste collection and disposal needed to be studied. At about the same time the world was beginning to become united on the climate change front and the important role of recycling and of informal recyclers was starting to gain legitimacy in Brazil as well as in Colombia. The labyrinthine laws and decrees that the Colombian national government and the Bogotá city government enacted and enforced over the next decade (as noted above), however, would appear to contradict the single-mindedness of their sustainability goals and of the city's recycling programs within that national legal context.

While there have been recent pronouncements about embracing recycling as one of the principal strategic responses to global warming by leadership in Latin America's cities (UN Habitat, 2010; Waste Pickers without Frontiers, 2008), public policy's record on solid waste management systems in the hemisphere make this call to arms welcome but disingenuous. Colombia's lackluster progress in leveraging the informal recycling sector into sustainable, municipal solid waste management schemes until very recently belies this intent, as does the fact that Doña Juana remains today Bogotá's only landfill in use. It has been calculated that the majority of the waste that arrives today at Doña Juana could be recycled if more efficient recycling programs were in effect (Parra, 2007; Ruiz, 2011). These pronouncements render the effects of partnerships like PP even more impressive. By making its incremental dent in the amount of waste that reaches the gates of the city's solo garbage dump, it provides a real-world example of how to reduce the inherent risks present at all landfills in underregulated developing countries, especially those at large, unwieldy dumps like Doña Juana (Gutberlet 2000; Martinez, 2010; UN Habitat, 2010). Perhaps most important, *Proyecto Pensilvania* offers a model of private-civil sector collaboration that is worthy of consideration in other industries and countries.

Concluding Remarks

Bogotá's checkered legal trajectory as it moved toward including recyclers and the Bogotá Association of Recyclers into a sustainable solid waste management agenda has been inconsistent and has sometimes resulted in negative externalities for the principal stakeholders (e.g., confusing regulations for businesses, for tax-paying residents of Bogotá, and for the informal recyclers themselves). It has also resulted, at times, in trash collection and disposal inefficiencies and in the undermining of nation- and citywide sustainability goals. Nevertheless, this chapter has demonstrated that in spite of legal setbacks, the indefatigable efforts of the Association of Recyclers of Bogotá gave birth to a unique collaborative partnership with socially responsible private enterprises. Could a PP work in other locales?

Several prerequisites are needed for a PP-type initiative to be imitated successfully in other developing country cities. First, landfills where waste-pickers openly scavenge through trash and are exposed to many health and environmental dangers should be closed and replaced by "sanitary" or closed landfills. More important, since this change from open to closed landfills almost always involves displacing the waste-pickers whose livelihoods depend on their daily collection of potentially recyclable goods, such closings offer an opportune moment for governments to recognize waste-pickers' contribution to the city's recycling targets, and for waste-pickers to unite and organize themselves. In fact, the history of Bogotá's recycling associations is linked to the dismantling of the open garbage dumps that, once closed, spurred the desperate waste-pickers from around the city to join forces, form cooperatives, and attempt to achieve alternative sources of employment with a unified voice.

The second requirement for potential private sector and recycling partnerships dictates that a city's informal recyclers be organized into a formal organizational structure, such as an association or a cooperative, that fulfills all legal requirements for an entity to operate and compete. Such a structure would include elected leaders with clear authority lines, a formal mandate for defining the association's objectives, and strategies for achieving them. According to Nohra Padilla, the current president of Bogotá's ARB, the formation of associations has been "profoundly important" in laying the foundation for social inclusion, and that "[e]ach group (of waste-pickers) by itself is fragile and vulnerable, but united we have more weight and are more capable of carrying on an organized struggle" (Samson, 2009a:37).

A third crucial component that a developing country needs to have in place if partnerships like PP are to thrive is an adequate regulatory regime. The evidence suggests that legalizing waste-picking and informal recycling activities at the national public-policy level is a sine qua non toward bettering the recyclers' lives and toward achieving efficient and integrative solid waste management systems that contribute to reducing the risks of climate change (Medina, 2008b). That said, having the appropriate laws on the books is a necessary but not sufficient precondition. What is key to any municipal solid waste management partnership, whether it be private-civil like PP, or civil-public, like the other partnerships that the ARB has formed with Bogotá's government, is that the legislation be consistent at the national and municipal levels, transparent, easily interpreted, and enforceable. Just as the tenets of institutional theory predict, if what is on paper—a government decree, law, regulation, or amendment—cannot be enforced, then there is little likelihood that the law will be upheld (North, 1990).

The legacy of the PP is not so much what it is but what it suggests about the recycling capability and the tenacity of a city's informal recyclers. Despite public policy setbacks, social marginalization, extreme poverty, a two-decade struggle, social cleansing aimed at poor populations, and seemingly insurmountable odds, the Bogotá Association of Recyclers has managed to survive and flourish. The author joins other scholars and practitioners in concluding that not only are Bogota's and Colombia's recycling and sustainability achievements enhanced by the ARB's existence, but, that an integrated municipal solid waste management system, one that incorporates local informal recyclers and their grass-roots associations into its formal business model, is one of the smartest ways for Bogotá and other developing country cities to grow in a sustainable way (Birkbeck, 1978; Breslin, 2002; De Soto: 1989; Easterly, 2006; *The Economist*, 2009; Hoornweg and Giannelli, 2007). It is smart for a globally warming planet. It is smart for the elected leadership who can take credit for improved recycling rates. It is smart for municipal solid waste management system efficiency. It is smart for firms looking for socially responsible partnerships. And it is smart for society to recognize the environmental contribution of informal recyclers and to welcome them and their expertise into the twenty-first century.

References

Birkbeck, C. 1978. Self-employed proletarians in an informal factory: The case of Cali's garbage dump. *World Development* 6(9–10): 1173–1185.

Breslin, P. 2002. Bogotá's recyclers find a niche—and respect. *Grassroots Development* 23(1): 26–28.

CEMPRE. 2010. http://www.cempre.org (Accessed February 21, 2010).

Cointreau, S. 2006. *Occupational and environmental health issues of solid waste management: Special emphasis on middle- and lower-income countries.* The International Bank for Reconstruction and Development/The World Bank. http://www.wds.worldbank.org/external/default/WDSContentServer/WDSP/IB/2007/07/03/000020953_20070703143901/Rendered/PDF/337790REVISED0up1201PUBLIC1.pdf (Accessed September 1, 2010).

De Soto, H. 1989. *The other path.* New York: Basic books.

Dias, S. M. 2000. Integrating waste pickers for sustainable recycling. In *Planning for Integrated Solid Waste Management—Collaborative Working Group Workshop*, 18–21 September, Manila, the Philippines. http://www.docstoc.com/docs/DownloadDoc.aspx?doc_id=20395589 (Accessed February 2, 2011).

Easterly, W. 2006. *The white man's burden: Why the west's efforts to aid the rest have done so much ill and so little good.* New York: Penguin Press.

(The) *Economist.* 2009. Muck and brass plates: Waste disposal in Colombia. U.S. Edition, 13 June. http://www.lexisnexis.com/us/lnacademic/results/docview/docview.do?docLinkInd=true&risb=21_T7199681467&format=GNBFI&sort=BOOLEAN&startDocNo=1&resultsUrlKey=29_T7199681470&cisb=22_T7199681469&treeMax=true&treeWidth=0&csi=7955&docNo=1 (Accessed November 23, 2010).

EPA. 2009. Municipal solid waste generation, recycling, and disposal in the United States: Facts and figures for 2009. http://www.epa.gov/waste/nonhaz/municipal/pubs/msw2009-fs.pdf (Accessed March 15, 2011).

———. 2011. *Landfills.* http://www.epa.gov/waste/nonhaz/municipal/landfill.htm (Accessed January 22, 2011).

Estrada, D. 2009. Latin America: The climate clock is ticking. *Tierramérica online.* http://www.tierramerica.info/nota.php?lang=eng&idnews=3277&olt=454 (Accessed January 28, 2010).

Ethical Corporation. 2008. http://www.ethicalcorp.com/communications-reporting/natura-%E2%80%93-natural-success-brazil (Accessed June 17, 2011).

Godoy, E. 2010. Los recicladores levantan su bandera en Cancun. [Recyclers raise their flag in Cancun]. http://ipsnoticias.net/pring.asp?idnews=97052 (Accessed January 21, 2011).

Guillermoprieto, A. 1990. Letter from Mexico City. *The New Yorker,* September 17: 93–104.

Gutberlet, J. 2000. Sustainability: A new paradigm for industrial production. *International Journal of Sustainability in Higher Education* 1(3): 225–236.

Hendron, D. 2006. Large landslide risks in solid waste facilities: Geotechnical fundamentals count. *Geo-Strata,* March–April: 28–30.

Hoornweg, D. and N. Giannelli. 2007. Managing municipal solid waste in Latin America and the Caribbean. *Gridlines* 28(October): 1–4.

Johannessen L., and G. Boyer. 1999. *Observations of solid waste landfills in developing countries: Africa, Asia, and Latin America.* The International Bank for Reconstruction and Development. http://www.worldbank.org/urban/solid_wm/erm/CWG%20folder/uwp3.pdf (Accessed February 15, 2011).

Martinez, C. 2010. Informal waste-pickers in Latin America: Sustainable and equitable solutions in the dumps. In *Global sustainability as a business imperative,* ed. J. Stoner and C. Wankel, Chapter 12, 199–217. New York: Palgrave Sustainability Series.

———. 2011. Personal conversations with the leadership of the ARB and other informed sources in Bogotá, Colombia, January 2011.

Medina, M. 2005. Serving the unserved: Informal refuse collection in Mexico. *Waste Management & Research* 23(5): 390–397.

———. 2007. *The world's scavengers: Salvaging for sustainable consumption and production.* Lanham, MD: AltaMira Press.

———. 2008a. The informal recycling sector in developing countries. *Gridlines,* 44 (1–4).

———. 2008b. Commmunity-based recycling initiatives. *Grassroots Development,* 29(1): 26–31.

———. 2009. Global recycling supply chains and waste picking in developing countries. *World Institute for Development Economics Research,* United Nations University. http://www.wider.unu.edu/publications/newsletter/articles/en_GB/12-2009-wider-angle-1/ (Accessed January 28, 2011).

Nas, P., and J. Rivke. 2004. Informal waste management. *Environment, Development and Sustainability* 6: 337–353.

North, D. 1990. *Institutions, institutional change and economic performance.* New York: Cambridge University Press.

Parra, F. 2007. Reciclaje popular y politicas publicas sobre manejo de residuos en Bogotá (Colombia). In *Recicloscopio,* ed. P. Schamber and F. Suarez, 63–82. Buenos Aires: Prometeo Libros.

PDDF (Project Design Document Form). 2009. *UNFCCC Project: Dona Juana landfill gas-to-energy project.* http://cdm.unfccc.int/filestorage/ (Accessed March 20, 2011).

Project design document form, 2006. UNFCCC. http://cdm.unfccc.int/filestorage/V/C/3/VC3GPFAZAC9FORF1G4EXGWJS39DA8P/CDM_PDD_Salvador_V3.4.doc (Accessed February15, 2010).

Ruiz, S. 2011. Information provided in face-to-face conversation with author, January.

Samson, M. 2009a. Asociación de recicladores de Bogotá (ARB), Colombia, [Association of Recyclers of Bogotá]. In *Refusing to be cast aside: Waste pickers organizing around the orld,* ed. M. Samson 36–43. Cambridge, MA: Women in Informal Employment: Globalizing and Organizing (WIEGO).

———. 2009b. Colombia—El dere,cho a competir y los derechoshumanos [Colombia—The right to compete and human rights). In *Refusing to be cast aside: Waste pickers organizing around the world,* ed. M. Samson, 73–76. Cambridge, MA: Women in Informal Employment: Globalizing and Organizing (WIEGO).

Schamber, P., and F. Suarez. 2007. *Recicloscopio.* Buenos Aires: Prometeo Libros.

Tellus Institute, 2008. *Assessment of materials management options for the Massachusetts solid waste master plan review: Final report appendices.* http://www.mass.gov/dep/recycle/priorities/tellusmma.pdf (Accessed July 31, 2011).

Tokyo 3R Statement. 2009. http://www.uncrd.or.jp/env/spc/docs/Tokyo-3R-Statement-12Nov2009.pdf (Accessed February 10, 2011).

Tuck, L. 2009. *Latin America's green path forward.* Tierramérica online. http://www.tierramerica.info/nota.php?lang=eng&idnews=3282&olt=455 (Accessed March 15, 2011).

UNFCC. 2011. *Status of ratification of the Kyoto Protocol.* http://unfccc.int/kyoto_protocol/status_of_ratification/items/2613.php (Accessed April 12, 2011).

UN Habitat. 2010. *Solid waste management in the world's cities.* Malta: Gutenberg Press.

Waste Pickers without Frontiers. 2008. Conference proceedings from the first international and third Latin American conference of waste-pickers. http://www.wiego.org/reports/WastePickers-2008.pdf (Accessed February 20, 2010).

World Economic Forum. 2007. *LatinAmerica@Risk.* http://www.scribd.com/doc/6310139/Latin-America-Risk (Accessed November 20, 2010).

PART III

Going Forward—Emerging Questions and Desirable Evolution in Global Sustainability Risk Management

CHAPTER 9

Managing the Risks of Carbon Sequestration: Liability Concerns and Alternatives

Elizabeth Lokey Aldrich, Cassandra Koerner, Joseph C. Perkowski, and Travis L. McLing

Introduction

Business managers involved in the production of greenhouse gases have several critical concerns. In parallel with efforts to improve performance, reduce deployment schedules, and increase cost-effectiveness, those responsible for carbon sequestration solutions must address their responsibility for long-term liability. Existing regulations for subsurface storage of petroleum and natural gas provides some guidance for carbon dioxide (CO_2) storage projects. Also, collective industry experience both on similar projects and on incipient sequestration efforts can give us guidance on the range of procedures that need to be managed. However, at present we lack a comprehensive regulatory framework to enable the effective management of long-term sequestration project operations.

The United States uses coal for nearly 50 percent of its electricity production (U.S. Energy Information Administration, 2010a), and compared to other fossil fuels, coal releases more CO_2 when combusted (U.S. Energy Information Administration, 2010b). Due to the warming potential of released CO_2, improved use of coal and sequestration of the CO_2 from combustion is being pursued avidly with public and private research. Carbon capture and sequestration (CCS), which involves the separation of CO_2 from stack emissions and geologic sequestration of CO_2, is a promising approach to address emissions from both natural gas and coal-fired power plants. In this chapter we use "sequestration" in lieu of "storage" to describe the permanent containment of CO_2 in geologic reservoirs since CCS has a more enduring connotation.

CCS could prevent the release of up to 90 percent of CO_2 emissions at a stationary power plant or industrial facility if CO_2 were captured directly at the source, separated from other by-products to a mandatory minimum purity level, transported to a geologic formation, and stored for an indefinite period of time (Pacala and Socolow, 2004). The geologic formations conducive to CCS are often areas where coal, oil, or gas have been extracted or where saline water resides in the pore spaces of certain rock formations, thousands of feet underground. CO_2 is "indirectly" in a sense sequestered where it is injected to repressurize oil (or, in some more recent applications, natural gas) fields to increase production. During this process, known as enhanced oil recovery (EOR) or enhanced gas recovery (EGR), approximately 50 percent to 60 percent of the CO_2 injected remains in the ground (International Energy Agency Greenhouse Gas R&D Programme, 2010), with the exact amount heavily dependent on the capacity of the formation to hold CO_2. CCS differs from EOR because the goal of CCS is to maintain the CO_2 in the ground and maximize reservoir storage potential.

Key questions require a set of answers that are yet to be developed. Who is liable for CO_2 leakage long after the establishment of the injection sites? Who is responsible for the ongoing verification effort when CO_2 injection ceases? Since it is possible in some configurations that CO_2 may inadvertently migrate into areas containing resources such as drinkable water, how does a business manage overall liability (including the possible loss of otherwise recoverable resources)? When the geographic extent of storage sites crosses multiple jurisdictions, how do we resolve liability involving multiple stakeholders?

Geological sequestration involves long timescales, perhaps as long as hundreds or even thousands of years. Since it is unreasonable to expect private sector liability and responsibility to remain directly with any business owner over such a long timescale, a fully public or else at least a public/private arrangement is inevitable. Several efforts are currently underway to examine these issues in depth. Proposals that have been suggested range from relatively simple extensions of current subsurface regulations to more elaborate long-range stewardship obligations involving multiple layers of government. This chapter examines the risks associated with carbon capture and handling the risks and liabilities, by building on legacy legal arrangements for liability management. It then summarizes questions that remain unanswered from an institutional perspective.

There are three major constraints on our analysis. The first is that information is continually being compiled, and new important suggestions for alternate policy approaches are being developed. Just one example is the very recent Senate testimony regarding proposed bill S-699, held on May 12, 2011, and available from the website of the Senate Committee on Energy and Natural Resources. The second is that neither cost calculations for establishing the basic sequestration infrastructure, nor the additional costs for obtaining liability protection, are well established. The third is that the sheer anticipated volume of sequestration needed to reduce global warming significantly requires careful and continuous attention to liability concerns well beyond past experience with indirectly analogous subsurface industrial operations. Therefore, our presentation is intended to assist the interested business manager in understanding the basics of the current dialogue on CCS liability as another element of necessary risk management, and in this way serve as a start toward creating a specific investment strategy.

Risks of Carbon Sequestration

There are some risks to human and environmental health associated with carbon injection. Many anti-CCS activists have pointed to the dangers of CCS by comparing it to incidents of natural CO_2 release from lakes near volcanoes. In 1986, Lake Nyos, located in northwestern Cameroon, released a lethal amount of CO_2 suffocating over 1,700 people in its wake and leaving hundreds of others injured (Barberi, Chelini, Marinelli, and Martin, 1989). As dramatic and tragic as this example is, it is highly unlikely that if carbon were to escape from injection spaces, similar issues would occur because scientists expect the carbon will be held in pore spaces and little will be released at one time (Klass and Wilson, 2009). The Intergovernmental Panel on Climate Change (IPCC, 2005) explains that these risks are already being managed by the oil and gas industry on a regular basis via monitoring, using controls that are in place. CO_2 does, however, present risks to plant life by raising the CO_2 levels in the soil. Large releases of CO_2 in the soil can be lethal to plants and animals (Klass and Wilson, 2009). Human health can also be at risk from the effects of CO_2 on the subsurface where the injection displaces brine or mixes with and degrades the quality of subsurface water reservoirs. Subsurface aquifers are highly regulated and will be monitored carefully through safeguards put in place by the new Environmental Protection Agency (EPA) Underground Injection Control Class VI well rule.

In addition to these risks, maintaining the stability of the subsurface looms as a large risk that must be managed. Choosing an appropriate site for injecting CO_2 can protect the injectors and regulators from a number of subsurface issues such as breaching of the caprock, ground inflation, and induced seismicity. Researchers reported increased seismicity in the Denver area following the injection of waste fluids into subsurface spaces (Healy, Rubey, Griggs, and Raleigh, 1968), and more recently concerns have erupted in Arkansas over hydraulic-fracturing (fracking), which seems to be having a similar seismic impact in shale-laden areas (Eddington, 2011). These risks have been addressed previously in some areas such as Ohio (Sminchak, Gupta, Byrer, et al., 2002). Findings show that injection into more porous formations is less likely to experience fracturing, seismic activity, and breaches to caprock integrity than less porous geologic features (Sminchak, Gupta, Byrer, et al., 2002). The In Salah CCS site in Algeria uses interferometric synthetic aperture radar (In SAR) technology to monitor surface changes on its site. Preliminary data show a small amount of bulging around injection wells and other surface deformation, which could be attributed to the injection of carbon nearly two kilometers below the surface (Knott, 2008).

The Nature of Sequestration Liability Risks

General

Recent allegations against Cenovus, the primary carbon injector at the Weyburn site in Saskatchewan, indicate that the site has sprung a leak and is threatening health and human safety on lands surrounding the oil field/injection site (Petro-Find Geochem, 2010). The Weyburn situation is the first major allegation of leaked CO_2 in a large-scale carbon injection situation and may highlight the need for greater attention to the risks and liability associated with CCS.

The prime motive for liability management is the protection of human health and environmental integrity, as well as protecting the owner from unintended consequences. Figure 9.1 illustrates the three main stages of CCS; each time period has its own associated risks. The time of preinjection and operation occurs over a period of approximately 20–30 years (Klass and Wilson, 2008). These risks can be covered with established insurance products although the closure phase, when liability can be transferred to the state, is defined differently in different jurisdictions. In Montana the waiting period before the state assumes responsibility is 30 years, in Australia the time period is 15 years, and in the EU it is 20 years (Jacobs and Stump, 2010). Given these relatively short timeframes, short-term liability instruments are most appropriate for covering risks in the

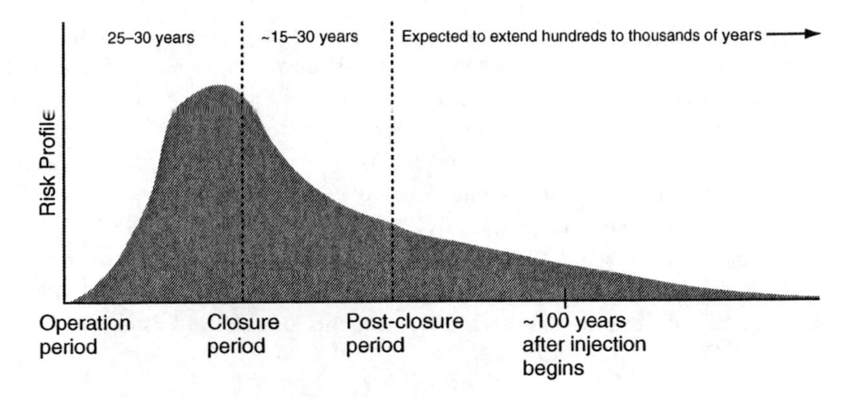

Figure 9.1 Timing and severity of CCS risks.
Source: Adapted from Wilson, 2005.

operating and closure periods. The postclosure period necessitates long-term liability since it spans an indefinite period of time, beginning at the end of the "closure" stage.

Short-Term Liability Risks: Preinjection, Operation, and Closure

One of the most important periods for determining liability occurs before injection begins. The size, porosity, and location of a geologic formation will determine the economic viability of the site as well as the amount of time a site can safely be operated (Pollak, Gresham, McCoy, et al., 2010). Establishing a site can come at a great cost to operators, depending on such factors as the response of the local population or the amount of pipeline and rights needed to move the CO_2 to the site.

Many of the short-term liability concerns associated with carbon sequestration have been based on decades of experience injecting carbon for EOR purposes (US DOE, 2011). Some expected operational liabilities include injury to workers, difficulties negotiating access to storage/access rights, poor well construction, poor siting of wells in areas of low injectivity, surface and subsurface property damage, induced seismicity, groundwater contamination, environmental damage, damage to the confinement zone, and atmospheric release of CO_2 (National Energy Technology Laboratory, 2009). The accumulated experience suggests that, as time passes, these risks will decrease.

Site closure requires additional financial assurances to guarantee well plugging and monitoring (Texas, 2010). Well plugging and abandonment plans are created during project development stages and have different requirements based on the location and type of well. Many abandonment plans require reclamation at and around the site to prevent environmental impacts (Fesmire, Rankin, Brooks, et al., 2007). Monitoring the integrity of wells and the formation is of utmost importance during the closure phase. In addition to any faults or fractures in the geologic formation, improperly constructed wells could provide a migration pathway for CO_2 (Federal Requirements under the UIC Program for CO_2 GS Wells, 2010). Subsurface migration of CO_2 and later mixing with water to leach minerals such as arsenic or hard metals into drinking waters is an additional risk during the closure phase (Federal Requirements under the UIC Program for CO_2 GS Wells, 2010).

Long-Term Liability Risks: Postclosure

Although short-term subsurface CCS liabilities are similar to those in EOR, the long-term implications require much more analysis because of their complexity. One major difference between EOR and CCS is the volume of CO_2 being injected into a formation. The increased volume and pressurization anticipated by CCS requires a certain type of stable geologic formation if operators intend on sequestering a great volume at a high pressure over thousands of years. Geologic formations where CCS is injected will be deep in the subsurface and covered by one or more impervious layers of rock to avoid leakage. The structure of the formation should have adequate porosity and permeability to hold CO_2. An additional valuable formation feature would be the existence of unpotable water that could potentially react with the CO_2 and could improve the affixation of the CO_2 within the formation (Office of Indian Energy and Economic Development, 2011).

One of the primary concerns in the postclosure phases is maintaining the CO_2 plume in the allotted space for centuries. The IPCC notes in its 2005 Special Report on Carbon Capture and Storage that there is little concern over seepage at well-selected sites and even minimal releases will be acceptable so long as they do not affect human and environmental health. The IPCC (2005, p. 34) reported that CO_2 storage in an "appropriately selected and managed reservoir" is likely to continue to contain 99 percent of its original volume over 1,000 years. The longer CO_2 remains in the subsurface, the more likely it is to combine with other minerals and be absorbed into either the rock or water (Federal Requirements under the UIC Program for CO_2 GS Wells, 2010). During all stages of CCS, monitoring of pressure, wells, and geology should be the operator's primary focus to reduce risk and associated damages.

Instruments for Sequestration Liability Management

Investment and liability risks are tied to the volume of gas being injected, its composition, and structure of the subsurface space that the gas is being injected into. Financial assurance is a necessary mechanism for successful CCS, but could potentially hold up development if the rules are too stringent. Alternatively, CCS could damage the environment and underground sources of drinking water (USDW) if the rules for injection and financial mechanisms to cover damages are not restrictive enough. Many have suggested that the most likely assurance program will be provided by operators over the short term and some public sector over the long term (Klass and Wilson, 2008). Because different liability instruments will be better suited for some phases of injection, the mechanisms are discussed separately.

Short-Term Liability Instruments

While the risk for damage during operations and shortly thereafter during closure may be great, this risk can most likely be quantified based on the area of injection, the volume of the injectate, and the proximity to populated or otherwise sensitive areas. Therefore, some insurance options are available during injection and closure that are not appropriate for coverage of long-term liability. The financial instruments to cover the operational phase of CCS, which begins with locating a suitable injection site and ends with well closure, include various forms of bondings and these require some upfront commitment to pay for costs of future potential damages. Typically these come in the form of trust funds, escrow accounts, letters of credit, and performance bonds. They are most applicable to short-term time frames since longer-term damages can be hard to assess and predict. In addition, financial institutions that would back up these financial commitments may not be in existence in the long term, and the long time frames may require capital for remediation and payment of potential damages to be tied up and unavailable for use by the insurers. Alternatively, private insurance can be used to cover short-term risks (Klass and Wilson, 2008). However, following the same line of reasoning, private insurance companies may not exist long enough to provide postclosure coverage.

[1] Trust Fund

A trust fund involves a third-party trustee that holds funds sufficient to cover estimated costs of damages (Federal Requirements under the UIC Program for CO_2 GS Wells, 2010). This approach is the surest form of funding but can be seen to put a burden on the operator's cash flow and often is based on the lowest estimates of damage (Texas, 2010).

[2] Escrow Account

An escrow account is a trust held by a third party that develops over time with a defined period of pay-ins, compared to having the fund financed and available at the start of the coverage period. These funds would have to be segregated from all other accounts and appropriated for the financial obligations caused from damage to human health or the environment. This mechanism could entail a higher risk than a Trust Fund if the operator stopped paying into the account or if there was a significant draw on the account early in development (Texas, 2010).

[3] Letter of Credit

A letter of credit (similar to payment bonds) is funding provided through a financial institution that can guarantee financial responsibility will be met if there is a demand for it. Most letters of credit are created based on estimated costs and do not require operators to pay up front; however, the operator must pay fees of 1–3 percent per year (Texas, 2010; Federal Requirements under the UIC Program for CO_2 GS Wells, 2010).

[4] Performance Bonds

Performance bonds allow for a financier or owner of a CCS project to be guaranteed compensation for monetary loss if the contractor who is conducting the operations of the CCS project fails to construct the project properly (Texas, 2010; Federal Requirements under the UIC Program for CO_2 GS Wells, 2010).

[5] Private Insurance

Some private insurance companies like SwissRe and ACE have begun exploring insurance for certain aspects of CCS like the injection and closure stage. Zurich Financial Services Group (2009) even provides an offering for this type of insurance. However, other private insurance companies are reluctant to enter this field as they are having a difficult time setting premiums because they do not have a good model to estimate CCS risk.

Private insurance differs from other financial mechanisms: since it is a cost reimbursement agreement, no company provides an automatic payout upon claim, but rather the provider may challenge the claim based on terminology used in the contract. These complex contracts would be different for each project, making it difficult to determine the likelihood of reimbursement for any foreseeable event. Additionally, because of the long-term nature of the postclosure stage and the expressed concern over the costliness of insurance required for monitoring and remediation over long-time horizons, private insurance may not be seen as a preferred financial assurance mechanism for long-term liability (Texas, 2010; Federal Requirements under the UIC Program for CO_2 GS Wells, 2010). Insurance could however be beneficial in shorter time frames, such as injection and closure phases. Most insurance companies are likely to provide only short-term offerings because they cannot guarantee their existence and accountability for the long-term contracts that would be required in the postclosure phase of injection.

Long-Term Liability Instruments

Types of long-term liability schemes that exist include comprehensive liability funds; funds that cover only monitoring and remediation; total indemnification of a site by the state or federal government; approaches to aggregate operational responsibilities encompassing multiple operators; and also possibly a three-tiered scheme that blends personal insurance with industry pooling and federal indemnification. Other options for insurance schemes used to cover catastrophic events such as floods, hazardous waste, and oil pipelines will not be covered because they are already covered in an EPA report and are not perfect corollaries for CCS (United States Environmental Protection Agency, 2008).

Industry Pooling

Some experts have suggested the government (or a private entity with government oversight) create a fund that multiple operators would pay into to cover liability rather than use private insurance. This method would pool risk for multiple operations through a required contribution or fee program that would likely be based on the project size and activity expectations (metric tons of CO_2 injected). Having operators pay into this fund during operations could promote responsible action if the amount being paid were tied to risk assessments (Interagency Task Force on Carbon Capture and Storage, 2010).

Federal compensation systems building on basic pooling mechanisms have also been described by several authors (Klass and Wilson, 2008; Interagency Task Force on Carbon Capture and Storage, 2010). This program could be modeled on a portion of the Price-Anderson Nuclear Industries Indemnity Act of 1957, or on the 1980 CERCLA provisions (U.S. Environmental Protection Agency, 2008). Under CERCLA, strict, joint, and several liabilities, which could make any of the parties participating in the project potentially responsible for damages to the environment, apply. If none of these entities is found responsible or can be identified and pay, cleanup costs are provided by the fund from the Superfund tax collected from chemical and industrial industries. Alternatively, the industry-pooled fund could be based on the Oil Spill Liability Trust Fund (OSLTF), which is funded by a per-barrel tax on petroleum produced or imported in the US (Jacobs and Stump, 2010).

Application of this type of pooling mechanism for CCS would entail the collection of a charge per metric ton of CO_2 injected. Alternatively, the fund could be based on a fee collected from all major CO_2 emitters, but this method would involve taxing many entities that do not engage in CCS. Such a fee would more closely resemble the fee that on-shore petroleum producers pay into the OSLTF since only oil producers that use vessels or deepwater ports would take advantage of the fund (Jacobs and Stump, 2010). Due to the resistance that this type of fund may face, most approaches seem to contemplate the collection of fees on a per metric ton injected basis. Most of the proposed pooling plans with comprehensive coverage and state indemnification omit strict, joint, and several liability in order to encourage the CCS industry (Rankin, 2009).

Louisiana, Kansas, Wyoming, and Texas have created stewardship funds to cover limited long-term liability. These funds handle the long-term monitoring and liability of sites and require CCS operators to pay a certain amount per metric ton of CO_2 injected. Each state fund covers slightly different aspects of the long-term liability. The likely impact of these funds on the CCS industry is not yet known, but their incomplete nature could lead to litigation and a reluctance of prospective operators to participate.

In the United States, Federal jurisdiction exists to some extent, and within the United States, six states have created industry-pooled funds to cover monitoring of sites. Outside the United States, the EU has created a fund to cover liability into the future. These U.S. state schemes differ in coverage and approach, with some states creating a long-term stewardship fund that requires the state to assume only limited long-term liabilities. Other states created a stewardship fund but require the state to assume all long-term liability. In contrast to the states that have created liability schemes, some states like Oklahoma, Utah, Washington, and West Virginia have passed legislation creating state-level permitting for CCS, but these states have refrained from creating any schemes to cover liability.

Indemnification of Liability to Promote Early Adoption or Demonstration Projects

To promote CCS, some states have offered to provide indemnification of liability. In the United States, the states of Texas and Illinois both made such offers for FutureGen project participants as they competed for federal funding. Through legislation, the State of Texas also offered to transfer the property rights of the CO_2 from the operator to the state and then exempt the project from tort liability. Subsequently, the State of Illinois passed legislation that involved the transfer of title for the CO_2, the purchase of third-party insurance for long-term storage if available, and indemnification for the FutureGen operator (Klass and Wilson, 2008).

In Canada, the Provincial Government of Alberta has taken this concept even further by assuming long-term liability for *all* CCS sites (Brooymans, 2010). This blanket assumption of liability in Alberta may have seemed logical to Albertans since the Crown owns all subsurface minerals. This type of liability assumption in the United States, however, is likely to be offered for only a discreet number of demonstration projects in order to stimulate the industry.

Another Alternative: Three-Tiered Liability Approach

The EPA has considered a three-tiered CCS liability proposal, much like the Price-Anderson Nuclear Industries Indemnity Act of 1957 for nuclear generators. Price-Anderson was created in the late 1950s to provide incentives for private development of nuclear power while still providing adequate compensation to victims of accidents tied to the industry. The nuclear developers' concerns at that time were similar to those of CCS developers today: the available insurance offerings did not seem adequate to cover catastrophic events, and investors were wary of entering into contracts and begin deployment with potentially high financial risks (Indemnification and Limitation of Liability, 2010).

The Price-Anderson Act involves a shared liability between the operator and general public, which may be more palatable than full federal or state assumption of liability. Nuclear generators pay for insurance that is capped at $375 million total liability per facility. If an accident depletes the individual plant insurance (the first tier of insurance), then the pooled-industry fund that totals $12.2 billion is tapped. This Fund is collected after the accident from each of the operating plants. Each plant must pay $17.5 million per year up to a maximum of $119 million per incident into this Fund if needed (Indemnification and Limitation of Liability, 2010; Inflation Adjustment to the Price-Anderson Act, 2008). Finally, if this Fund is depleted, then the federal government assumes liability and indemnifies the operator (US Environmental Protection Agency, 2008).

This three-tiered approach places some responsibility on the individual operator and then shares responsibility with the industry as a whole. It has the advantage that it creates a type of peer pressure on both the operator and the industry as a whole to be responsible (Rankin, 2009). Of course it would have to be tailored for CCS, since the appropriate minimum amount of insurance purchased may be contingent upon the amount of CO_2 injected in the site. So, instead of requiring all facilities to purchase the mandated level of $375 million in insurance as occurs in the nuclear industry, perhaps for CCS the amount of insurance purchased could be directly proportional to the volume of injectate. Likewise, instead of having industry participants pay retroactively for accidents that occur, perhaps a fee per metric ton as have been enacted by state-level laws should be collected and the Fund could grow over time, even after CCS operators retire, providing on-going maintenance and remediation.

Differences between CCS and the nuclear industry create some pitfalls for this approach that need to be addressed. Nuclear facilities usually have a permitted operating life of 40 years (Webster and LeMense, 2009) and nuclear operators can therefore hold an insurance policy that covers this length of time. The "in perpetuity" nature of CCS sites poses an issue to the first tier of insurance as the operator may not be an appropriate entity, since both it and its insurance plan will likely not persist "in perpetuity."

The second drawback to this approach is that the buildup of an industry fund to cover disasters may not be achieved in the short term and is highly dependent on how many CCS projects are created. Therefore, this three-tiered approach will most likely resemble a two-tiered approach in the short term as the federal government would have to provide indemnification after private insurance fails. This drawback also applies to the aforementioned comprehensive liability funds in Montana and North Dakota.

Since continuing critique of alternative liability management schemes will be essential before one universal formula can be developed (if indeed that can ever happen), and since a number of reviewers have spent considerable professional effort on this subject, a global summary of all policy and management ideas for sequestration liability management is not provided in this chapter. However, a flavor of the various perspectives is offered by the work of Patton and Trabucchi (2008), which in turn was examined by a team analysis headed by Carnegie Mellon University (CCSReg Project, 2009). Patton and Trabucchi question whether the nuclear liability model is complete since it focuses on damage compensation versus long-term continuing operational costs, and also on the basis that too much financial burden may be placed on the public in the longer term. The CMU report offers suggestions for establishing risk management in a way that keeps the burden on the project owners and developers rather than the general public by, for example, including the maintenance of the highest possible industrial standards during site screening; applying a standard of negligence to any tort liability during site operations;

enforcing sufficient financing of a long-term "sinking fund" that minimizes long-term public sector financial exposure; and establishing a sustainable approach to adjusting CO_2 credit trading valuations if CO_2 leakage occurs in the later years.

Aggregation of Operational Responsibility for Liability

If sequestration reaches the volume level of a major nationwide activity, and if (as part of this growth in business) there become many instances of multiple source entities contributing CO_2 to a "group" (collective) site, the opportunity may exist for a joint-venture operator assuming all management risks including liability in a manner superior to that which any individual operator could propose. In this way it is conceivable that coverage costs would be optimized (perhaps as an offset to otherwise unused tax credits or GHG trading credits, for example). Fundamental liability responsibility would not change but the manner in which it is managed could be substantially modified. This notion could in some fashion be analogous to the common practice of oil field "unitization" for optimizing conventional oil recovery in neighboring fields.

Other Ongoing and Prospective Efforts

The Interstate Oil and Gas Compact Commission (IOGCC) continues to improve its database regarding State liability regulatory activity and (as of this writing) has submitted a "Phase 3" proposal to the DOE for an in-depth analysis of long-term CCS "stewardship" issues. Providing an exhaustive list of related activities by other interested parties is not practical, but the activities of the relatively new Global CCS Institute (see www.globalcc sinstitute.com) may also be of interest to liability managers. The US DOE also continues to fund multiyear liability assessments through universities such as Princeton (see www.princeton.edu/pei/news/archive/?id=2892). Regarding public lands management, the Department of Interior's 2007 report is useful (U.S. Department of Interior, 2007).

Possible Comprehensive Liability Schemes

Given the dramatic difference in the amount of risk associated with each stage of CCS, different liability schemes should be examined and constructively critiqued during this early stage of possible full-scale deployment. For the operation and closure stage of CCS, a variety of products including private insurance, bonding, trust funds, escrow accounts, and other mechanisms may well be appropriate. All of these options are possible due to the limited time frame of injection and closure, which can be defined as anywhere between 5 and 50 years. Within this time frame there is reasonable assurance that the institutions providing the coverage will be in existence throughout the duration of the project, and private insurance companies and traditional financial instruments like escrow accounts, performance bonds, and letters of credit will be able to cover the risk. Corrective policies can also be put in place that provide recourse action if the institutions fail. The more interesting question arises when one considers what the best long-term liability schemes will (or should) be.

Criteria for Evaluating a Comprehensive Liability Scheme

Prior to analyzing all liability approaches and recommending any one liability scheme covering all stages of the CCS process, it is essential to identify the key features of a preferred comprehensive liability scheme. While these criteria are somewhat arbitrary, the following six points comprise what we deem most important for a CCS liability scheme. We should note that while they have been derived in a US context, some variations may be applicable overseas; a detailed overseas analysis would be beyond the scope of our effort at this point.

- *The scheme should allow for commercialization of CCS.* If the liability scheme chosen puts too much risk or financial burden on the operator, it is likely to stifle development. If CCS is deemed to be an important solution to climate change, then a goal of the liability approach selected should be to encourage development.
- *The approach selected should not overly burden the public* by placing all liability on federal or state government entities. A scheme of this type would create an unfair subsidy for the CCS industry and could lead to careless practices during injection.
- The scheme should *clearly identify liable parties* so as to minimize litigation, which could slow cleanup efforts and waste taxpayer dollars.
- *Coverage should be complete* and comprehensive, covering all damages so as to minimize litigation and lack of full remediation.
- The scheme chosen should *encourage operators and the industry as a whole to act responsibly* during operations and closure.
- The scheme should *provide funds readily*, i.e., whenever needed, so that delays in cleanup or payment for damages do not occur.

Analysis of Comprehensive Liability Schemes

Very few of the liability schemes that have thus far been implemented, with the exception of the schemes in Alberta, North Dakota, and Montana, cover long-term liability in a comprehensive way. As a result, these incomplete liability schemes are likely to cause confusion, misunderstandings, and litigation as they are implemented.

The EPA Class VI well rule has a limited scope of liability in that it covers only contaminated groundwater. Damage to vegetation, animals, or induced seismicity is not covered by this rule. And, given the recent allegations of leaked CO_2 at the Weyburn site in Canada, which have allegedly caused damage to the flora and fauna of the area, and reports of induced seismicity from injected water and natural gas in Colorado and Alabama, safeguards to protect more than just USDW are needed (Eddington, 2011; Johnson, 2009; Robertson, 2011). Furthermore, climate liability for escaped metric tons of CO_2, which may qualify as a stationary source of pollution under more stringent future EPA greenhouse gas regulations or comprehensive federal greenhouse gas legislation, is not addressed by this rule.

The Class VI rule is also flawed because it leaves much up to the discretion of the EPA Director. Whether or not operators and owners of the site are held liable for a period beyond the 50-year closure period is (at least currently) unclear. And, under this Class VI rule, these owners and operators are unable to transfer liability to a third party in the future, which will pose a huge threat to the industry since no owner or operator will want to face the threat of assuming permanent liability for the site. Furthermore, it is up to the discretion of the Director to determine the amount of funding and time necessary to insure the project (Federal Requirements under the UIC Program for CO_2 GS Wells, 2010). Both of these unknowns could stymie the CCS industry as developers may be hesitant to invest under these circumstances. In contrast to the EPA Class VI rule, European Union Directives on CCS allow the transfer of authority and responsibility for the site and allow for the collection of money for future monitoring from operators, creating a sustainable system for funding (Journal of the European Union, 2009; Directive 2003; Directive 2004).

The liability schemes implemented in Louisiana, Kansas, Wyoming, and Texas fail to provide coverage of civil and climate liabilities. While these states have made efforts to set up funds that cover the monitoring and remediation of CCS sites in the future, these funds do not cover civil or climate liabilities. And, these states have taken differing approaches to the amount of ownership of CO_2 and liability they assume. Louisiana has stated in its law that neither the owner nor the state is responsible for the CO_2

after closure (Louisiana HB 661 §1109(A) (2), 2009). Kansas's law does not mention transfer of liability or ownership of the CO_2 (Kansas HB 2419 §3, 2007). Wyoming claims that the state is neither liable for nor immune from suits for damages from the CO_2 in the long term (Wyoming HB 17 §1, 2010). Texas has assumed liability and ownership of the CO_2 from offshore applications, but is silent on this issue for onshore CCS operations (Texas SB 1796, 2009; Texas SB 1387, 2009).

The liability schemes implemented in Alberta, North Dakota, and Montana involve the state or province taking title to the CO_2 and assuming the long-term liability after a designated period of time. North Dakota and Montana have set up funds for long-term management. Both of these funds are based on a charge per metric ton of CO_2 injected (North Dakota SB 2095, 2009; Montana SB 498, 2009). These schemes are similar to an IOGCC proposal calling for a ten-year closure period followed by monitoring and remediation with a fund that is generated from payments made by CCS operators (IOGCC, 2007).

While these schemes may be successful at the state and provincial level, assumption of this much liability by the federal government or by other states may not be feasible or desirable. Essentially, the type of fund created by North Dakota and Montana and proposed by the IOGCC has two tiers. The first tier consists of an industry-generated fund with revenues from injectors. If funding from this tier is depleted, then the second tier, which allows the state to indemnify the project and assume all liability, is utilized. Until the first tier has accumulated funds to cover a disaster, this type of liability scheme essentially involves the state assuming all liability for the project and indemnifying it. Similar to full indemnification provided by Alberta, and originally offered for the FutureGen project, this arrangement could place a large burden on the public to help subsidize the industry.

These two-tiered proposals mimic a CERCLA postclosure liability fund that was imposed until the mid-1980s. Under this now defunct scheme, hazardous waste disposal operators could apply to have liability and future monitoring and maintenance costs covered by an industry-generated fund, capped at $200 million, after they had operated for five years and remained within the confines of their permits. This fund, however, was abandoned in the mid-80s because it was discovered that there was less incentive for hazardous waste disposal operators to pursue permanent management techniques when they knew that they would not be responsible for the site in the medium and long terms (Rankin, 2009). Likewise, the two-tiered state liability schemes that transfer all liability away from the operator too soon after operations or closure may not place enough responsibility on individual site managers and may also fail to provide adequate incentives for long-term stewardship of the site. Due to the drawbacks of a two-tiered liability structure, consideration of other schemes is necessary.

Considerations for a Homogenous US Comprehensive Liability Scheme

The three-tiered approach to liability may provide the best solution for CCS as it fulfills the criteria described above for a successful liability scheme. Requiring CCS operators to hold private insurance for a specified amount, which may be dependent on the amount of CO_2 injected, can provide proper levels of incentives to ensure that owners conduct their operations in safe and prudent ways. Of course, the problem of how to hold an entity responsible for private insurance far into the future becomes challenging. However, a prescribed number of years on the liability scheme could be defined, and operators could take out insurance for that period. Since the risks associated with a site taper off dramatically after the operations and closure phases are completed, it may be appropriate and necessary for the individual insurance plan to be abandoned when those phases are over. Given the nuclear industry's ability to obtain insurance for the initial 40 years that plants are permitted, a requirement of private insurance for at least 40 years of CCS project operations should be reasonable. Also, given insurance companies' early interest in insuring CCS operations for short-time horizons, the financial markets will most likely fill the need for this type of policy.

If an operator prefers an escrow account, letter of credit, or other financial instrument to back up its operations and closure, it is possible that this type of insurance could replace private insurance for this time period. However, it is important that some synchronization of expectations on the length and coverage of the insurance policy to cover the preoperation, operation, and closure time frame be made in order to ensure that the second tier of liability not be tapped into unnecessarily. The exact number of years that an operator must hold this private policy will be up for debate, but it seems likely that it would need to persist at least through preoperation, operation, and closure. It seems logical that the amount of coverage necessary be correlated to the amount of CO_2 injected, but the type of reservoir and its characteristics may also be important in determining coverage details.

The second tier of liability, a "shared pool," would cover large-scale disasters that deplete the first tier. Ongoing monitoring and remediation of wells would help to encourage industry participants to be prudent in their operations. Poor operation of a site would taint an operator or industry partner in their relationships with other prospective customers. The pitfall of this industry-pooled fund not being available until adequate funds have been paid into it is definitely a "chicken-and-egg" problem that can probably only be solved by collecting the tax for this fund from injectors during early CCS operations and relying on the federal assumption of liability (the third tier of the scheme) to cover damages that surpass the first tier. Alternatively, the fund could start with an initial minimum balance borrowed from (and then repaid to) the US Treasury (Jacobs and Stump, 2010). Since an operator will not pay into the fund indefinitely, use of this second tier to cover monitoring and remediation costs of old sites will necessitate the existence of future CCS operators who will pay into this fund.

Finally, the third tier of liability would ideally not need to be used. In the nuclear industry, the second tier of liability was utilized after the Three Mile Island meltdown in 1979 for payouts of $71 million, but the third tier has never been utilized, costing the US public nothing (Nuclear Energy Institute, 2010). If CCS proves at least as safe as the nuclear industry, this type of liability scheme should not overly burden the public through use of federal funds for cleanup costs that subsidize the industry. If such a scheme were in place for a number of years and then it was decided that CCS is not an appropriate solution to climate change, there might be no future CCS operators to pay into a fund that supports the monitoring and remediation of old sites and the government might then have to assume a financial role for this third tier of liability.

For this scheme to be implemented successfully, CCS operations would need to be explicitly exempt from environmental statutes like the RCRA, CERCLA, CAA, and the current Class VI rules under the Safe Drinking Water Act. If they were not exempt, they would be subject to these multiple liability regimes. With exemption from these statutes, protection against possible damages due to CCS (climate change, induced seismic activity, contamination of groundwater, impact on human health or health of flora and fauna) could be provided by some prudent and enforceable version of the three-tiered scheme.

Acknowledgments

The authors would like to thank Melisa Pollak, Research Fellow in the Science, Technology and Public Policy Program at the University of Minnesota Humphrey School of Public Affairs, for her insight into the state-level CCS liability legislation that is analyzed in this paper. We would also like to acknowledge the Center for Advanced Energy Studies' Energy Policy Institute for support of this research.

References

The IOGCC has, on its website, information regarding each state's regulatory progress. The website includes state-by-state summary information about the latest development of legal and regulatory infrastructure for the geologic storage of CO_2.

Space limitations constrain us from describing specific additional state rulings of interest to businesses on CCS liability, so further useful references are given below (Wyoming: HB 0057, HB 0058, HB 0080, HB 0089, HB 0090, HB 2860; Texas: HB 149; New Mexico: SB 145; Oklahoma: SB 610; North Dakota: SB 2139; Michigan: SB 775). Other very useful overviews are provided by de Figueiredo (2005), Ingelson, Kleffner, and Nielson (2010), and Aldrich and Koerner (2011).

Aldrich, E. L. and C. L. Koerner. 2011. Assessment of carbon capture and sequestration liability. *The Electricity Journal* 24(7): 35–48. doi:/10.1016/j.tej.2011.07.001.

Barberi, F., W. Chelini, G. Marinelli, M. Martin. 1989. The gas cloud of Lake Nyos (Cameroon, 1986): Results of the Italian technical mission. *Journal of Volcanology and Geothermal Research* 39(2–3): 125–134.

Brooymans, H. 2010, November 1. Alberta to assume liability for long-term carbon storage. *The Vancouver Sun.*

CCS Reg Project. 2009. *Carbon capture and sequestration: Framing the issues for regulation: An interim report from the CCS Reg Project.* Pittsburgh: Dept. of Engineering and Public Policy, Carnegie Mellon University.

de Figueiredo, M. A. 2005. *Property interests and liability of geologic carbon dioxide storage: A special report to the MIT Carbon Sequestration Initiative.* Cambridge, MA: MIT Laboratory for Energy and the Environment. http://sequestration.mit.edu/pdf/deFigueiredo_Property_Interests.pdf (Accessed July 31, 2011).

Directive 2003/87/EC, European Union. 2003. *Establishing a scheme for greenhouse gas emission allowance trading within the community and amending Council Directive 96/61/EC.* http://eur-lex.europa.eu/LexUriServ/LexUriServ.do?uri=OJ:L:2003:275:0032:0032:EN:PDF (Accessed July 31, 2011).

Directive 2004/35/CE, European Union. 2004. *On environmental liability with regard to the prevention and remedying of environmental damage.* http://eur-lex.europa.eu/LexUriServ/LexUriServ.do?uri=CELEX:32004L0035:EN:NOT (Accessed July 31, 2011).

Eddington, S. 2011, March 4. "Fracking" disposal sites suspended, likely linked to Arkansas earthquakes. *The Huffington Post.* http://www.huffingtonpost.com/2011/03/06/fracking-arkansas-earthquakes_n_831633.html (Accessed July 31, 2011).

Federal Requirements under the Underground Injection Control (UIC) Program for Carbon Dioxide (CO_2) Geologic Sequestration (GS) Wells, 75 Fed. Reg. 77230 2010 (to be codified at 40 C.F.R. pt. 124, 144, 145, 146, and 147).

Fesmire, M. E., A. Rankin, D. Brooks, and W. V. Jones. 2007. *A blueprint for the regulation of geologic sequestration of carbon dioxide in New Mexico.* New Mexico Energy, Minerals, Natural Resources Department, Oil Conservation Division. Retrieved from: http://www.gwpc.org/e-library/documents/co2/Mark%20Fesmire.pdf (Accessed July 31, 2011).

Healy, J. H., W. W. Rubey, D. T. Griggs, and C. B. Raleigh. 1968. The Denver earthquakes. *Science* 161(3848): 1301–1310.

H. B. 17 § 1, 60th Gen. Sess. WY. 2010. *Geologic sequestration special revenue account.* http://legisweb.state.wy.us/2010/Enroll/HB0017.pdf (Accessed July 31, 2011).

H. B. 0057, 60th Leg., Gen. Sess. WY. 2009. *Provides amendment to W.S. 34-1-152 (e) Ownership of pore space underlying surfaces.* http://legisweb.state.wy.us/2009/Enroll/HB0057.pdf (Accessed July 31, 2011).

H. B. 0058, 60th Leg., Gen. Sess. WY. 2009. *Relating to the responsibilities of injectors and pore space owners.* http://legisweb.state.wy.us/2009/Introduced/HB0058.pdf (Accessed July 31, 2011).

H. B. 0080, 60th Leg., Gen. Sess. WY. 2009. *Relating to the unitization of carbon sequestration sites.* http://legisweb.state.wy.us/2009/Bills/HB0080.pdf (Accessed July 31, 2011).

H. B. 0089, 59th Leg., Gen. Sess. WY. 2008. *Relating to the ownership of subsurface pore space.* http://legisweb.state.wy.us/2008/Introduced/HB0089.pdf (Accessed July 31, 2011).

H. B. 0090, 59th Leg., Gen. Sess. WY. 2008. *Relating to carbon sequestration, regulation of injection and appropriation.* http://legisweb.state.wy.us/2008/Enroll/HB0090.pdf (Accessed July 31, 2011).

H. B. 149, 78th Gen. Sess. TX. 2009. *Relating to the ownership of carbon dioxide captured by a clean coal project.* http://www.legis.state.tx.us/tlodocs/793/billtext/html/HB00149F.HTM (Accessed July 31, 2011).

H. B. 661 §1109 (A) (2) LA. 2009. *Conservation: Provides with respect to the geological sequestration of carbon dioxide, Cessation of storage operations; liability release.* http://www.legis.state.la.us/billdata/streamdocument.asp?did=659193 (Accessed July 31, 2011).

H. B. 2419 § 3 KS. 2007. *Carbon Dioxide Reduction Act.* http://www.kansas.gov/government/legislative/bills/2008/2419.pdf (Accessed July 31, 2011).

H. B. 2860 WV. 2009. *Relating to regulating the sequestration and storage of carbon dioxide.* http://www.legis.state.wv.us/Bill_Status/bills_text.cfm?billdoc=HB2860%20ENR%20SUB.htm&yr=2009&sesstype=RS&i=2860 (Accessed July 31, 2011).

Indemnification and Limitation of Liability, 42 U.S.C. § 2210. 2010. http://www.law.cornell.edu/uscode/42/usc_sec_42_00002210----000-.html (Accessed July 31, 2011).

Inflation Adjustment to the Price-Anderson Act Financial Protection Regulations, 73 Fed. Reg. 56451, 2008. (to be codified at 10 C.F.R. pt. 140).

Ingelson, A., A. Kleffner, and N. Nielson. 2010. Long term liability for carbon capture and storage in depleted North American oil and gas reservoirs: A comparative analysis. *Energy Law Journal* 31(2): 431–469.

Interagency Task Force on Carbon Capture and Storage. 2010. Report of the interagency task force on carbon capture and Storage. http://www.epa.gov/climatechange/policy/ccs_task_force.html (Accessed July 31, 2011).

Intergovernmental Panel on Climate Change (IPCC). 2005. *Special report on carbon dioxide capture and storage.* Cambridge, UK: Cambridge University Press.

International Energy Agency Greenhouse Gas R&D Programme. 2010. Weyburn Enhanced Oil Recovery Project. http://www.co2captureandstorage.info/project_specific.php?project_id=70 (Accessed July 31, 2011).

Interstate Oil & Gas Compact Commission. 2007. CO_2 storage: A legal and regulatory guide for states. http://iogcc.myshopify.com/collections/frontpage/products/co2-storage-a-legal-and-regulatory-guide-for-states-2008 (Accessed July 31, 2011).

Jacobs, W., and D. Stump. 2010. *Proposed liability framework for geological sequestration of carbon dioxide.* Cambridge, MA: Emmett Environmental Law and Policy Clinic, Harvard Law School. http://www.law.harvard.edu/programs/about/elp/ccswhitepaper.pdf (Accessed July 31, 2011).

Johnson, K. 2009, June 12. Quake zone: The natural gas industry's big fracking problem. *The Wall Street Journal—Environmental Capital Blog.* http://blogs.wsj.com/environmentalcapital/2009/06/12/quake-zone-the-natural-gas-industrys-big-fracking-problem/ (Accessed July 31, 2011).

Journal of the European Union. 2009. *Directive 2009/31/EC of the European Parliament and of the Council of 23 April 2009 on the geological storage of carbon dioxide and amending Council Directive 85/337/EEC,* European Parliament and Council Directives 2000/60/EC, 2001/80/EC, 2004/35/EC, 2006/12/EC, 2008/1/EC and Regulation (EC) No 1013/2006. http://eur-lex.europa.eu/LexUriServ/LexUriServ.do?uri=OJ:L:2009:140:0114:0135:EN:PDF (Accessed July 31, 2011).

Klass, A. B. and E. J. Wilson. 2008. Climate change and carbon sequestration: Assessing a liability regime for long-term storage of carbon dioxide. *Emory Law Journal* 58: 108–180.

———. 2009. Carbon capture and sequestration: Identifying and managing risks. *Issues in Legal Scholarship* 8(3). doi: 10.2202/1539-8323.1108.

Knott, T. 2008. Sealed under the Sahara. *Frontiers* 23: 18–25. http://www.bp.com/sectiongenericarticle.do?categoryId=9027098&contentId=7049642 (Accessed July 31, 2011).

National Energy Technology Laboratory. 2009. *Storage of captured carbon dioxide beneath federal lands.* DOE/NETL-2009/1358. http://www.netl.doe.gov/energy-analyses/pubs/Fed%20Land_403.01.02_050809.pdf (Accessed July 31, 2011).

Nuclear Energy Institute. 2010. *Price-Anderson Act provides effective liability insurance at no cost to the public.* http://nei.org/resourcesandstats/documentlibrary/safetyandsecurity/factsheet/priceandersonact/ (Accessed July 31, 2011).

Office of Indian Energy and Economic Development. 2011. *Geologic sequestration of carbon dioxide: Sequestration process and geologic reservoirs.* http://teeic.anl.gov/er/carbon/apptech/geoapp/index.cfm (Accessed July 31, 2011).

Pacala, S., and R. Socolow. 2004. Stabilization wedges: Solving the climate problem for the next 50 years with current technologies. *Science* 305: 968–972.

Patton, L. S., and C. Trabucchi. 2008. Storing carbon: Options for liability risk management, financial responsibility. *Daily Environment Report* 170: 1–22.

Petro-Find Geochem. 2010. Geochemical soil gas survey. *A site investigation of SW30-5-13-W2M. Weyburn Field, Saskatchewan.* http://www.ecojustice.ca/media-centre/media-release-files/petro-find-geochem-ltd.-report/at_download/file (Accessed July 31, 2011).

Pollak, M., R. L. Gresham, S. McCoy, and S. J. Phillips. 2010, May. State regulation of geologic sequestration: 2010 update. Lecture presented at Hilton Pittsburgh/David L. Lawrence Convention Center, Ninth Annual Conference on Carbon Capture and Sequestration, Pittsburgh, PA.

Rankin, A. G. 2009. Geologic sequestration of CO_2: How EPA's proposal falls short. *Natural Resources Journal* 49: 883–942.

Robertson, C. 2011, March 4. Waste wells to be closed in Arkansas. *The New York Times.* http://www.nytimes.com/2011/03/05/us/05fracking.html?_r=1&ref=earth (Accessed July 31, 2011).

S. B. 145, 49th Gen. Sess. NM. 2010. *Regarding ownership of pore space.* http://www.nmlegis.gov/lcs/_session.aspx?chamber=S&legtype=B&legno=%20%20%20145&year=10 (Accessed July 31, 2011).

S. B. 498, 61st Leg., Gen. Sess. MT. 2009. *Relating to permitting for carbon dioxide injection wells.* http://data.opi.mt.gov/bills/2009/BillPdf/SB0498.pdf (Accessed July 31, 2011).

S. B. 610, 52nd Gen. Sess. OK. 2009. *General storage of carbon dioxide act—Create.* http://www.statesurge.com/bills/sb610-oklahoma-571159 (Accessed July 31, 2011).

S. B. 775, 95th Leg., Gen. Sess. MI. 2009. *Regarding carbon sequestration fees and liability waivers.* http://www.legislature.mi.gov/documents/2009-2010/billintroduced/Senate/pdf/2009-SIB-0775.pdf (Accessed July 31, 2011).

S. B. 1387, 81st Gen. Sess. TX. 2009. *Relating to the implementation involving the capture, injection, sequestration, or geologic storage of carbon dioxide.* http://www.legis.state.tx.us/billlookup/Text.aspx?LegSess=81R&Bill=SB1387 (Accessed July 31, 2011).

S. B. 1796 §382.508 TX. 2009. *Texas health and safety code: Offshore geologic storage of carbon dioxide.* http://www.legis.state.tx.us/tlodocs/81R/billtext/html/HB01796F.HTM (Accessed July 31, 2011).

S. B. 2095, 61st Leg., Gen. Sess. ND. 2009. *Relating to the geologic storage of carbon dioxide.* http://www.legis.nd.gov/assembly/61-2009/bill-text/JQTA0300.pdf (Accessed July 31, 2011).

S. B. 2139, 61st Leg., Gen. Sess. ND. 2009. *Relating to the ownership of subsurface pore space.* http://www.legis.nd.gov/assembly/61-2009/bill-text/JQTB0400.pdf (Accessed July 31, 2011).

Sminchak, J., N. Gupta, C. Byrer, and P. Bergman. 2002. Issues related to seismic activity induced by the injection of CO_2 in deep saline aquifers. *Journal of Energy & Environmental Research* 2: 32–46. http://www.netl.doe.gov/publications/proceedings/01/carbon_seq/p37.pdf (Accessed July 31, 2011).

Texas. 2010. The Texas General Land Office, The Railroad Commission of Texas, The Texas Commission on Environmental Quality, The Bureau of Economic Geology. 2010. *Injection and geologic storage regulation of anthropogenic carbon dioxide*. http://www.rrc.state.tx.us/forms/reports/notices/SB1387-FinalReport.pdf (Accessed July 31, 2011).

United States Department of Energy. 2011. *Enhanced oil recovery/CO_2 injection*. http://www.fossil.energy.gov/programs/oilgas/eor/index.html (Accessed July 31, 2011).

United States Department of Interior. 2007. *Framework for geological carbon sequestration on public land*. http://groundwork.iogcc.org/sites/default/files/Framework%20for%20Geological%20Storage.pdf.

United States Energy Information Administration. 2010a. *Annual energy outlook with projections to 2035*. http://www.eia.doe.gov/oiaf/aeo/demand.html (Accessed July 31, 2011).

———. 2010b. *International energy outlook 2010*. Retrieved from: http://www.eia.doe.gov/oiaf/ieo/emissions.html (Accessed July 31, 2011).

United States Environmental Protection Agency. 2008. *Approaches to geologic sequestration site stewardship after site closure*. Office of Water. EPA 816-B-08-002. http://www.epa.gov/ogwdw000/uic/pdfs/support_uic_co$_2$_stewardshipforsiteclosure.pdf (Accessed July 31, 2011).

Webster, R., and J. LeMense. 2009. 40 Years and counting: Relicensing the first generation of nuclear power plants: Spotlight on safety at nuclear power plants: The view from Oyster Creek. *Pace Environmental Law Review* 26(2): 365–390.

Wilson, E. J. 2005. Subsurface property rights: Implications for geologic CO_2 sequestration. In *Underground injection science and technology, Volume 52 (Developments in water science)*, ed. C.-F. Tsang and J. A. Apps, 681–693. Boston: Elsevier Science.

Zurich Financial Services Group. 2009. *Zurich creates two new insurance policies to support greenhouse gas mitigation technologies, addressing the unique needs of carbon capture and sequestration*. http://www.businesswire.com/news/home/20090119005535/en/Zurich-creates-insurance-policies-support-green-house (Accessed July 31, 2011).

CHAPTER 10

Taking Actions to Deal with Climate Change Risks and Opportunities: Harnessing Superordinate Identities to Promote Knowledge Transfer and Creation

Aimée A. Kane and Alison L. Steele

Introduction

The reciprocal influence between climate change and corporations is undeniable. On the one hand, human actions have an irrefutable impact on the environment, one of the most dangerous examples being the correlation between increased carbon emissions and temperature extremes (Solomon, Plattner, Knutti, et al., 2009). As primary actors in the global economy, corporations represent a significant portion of human impact on the environment. Corporate carbon emissions, for example, result from activities including manufacturing, resource harvesting, business travel, and building occupancy (Environmental Information Administration, 2011). On the other hand, climate change influences corporations by increasing uncertainty, instability, and risk. Such risks include difficulties with crop production caused by temperature variation and fluctuations in water supply; shifts in regional weather patterns and heightened potential for natural disasters (Stern, 2007); and increased costs of heating, cooling, and maintaining air quality (Mickley, Jacob, and Field, 2004). The magnitude of this influence is so significant that the United States Securities and Exchange Commission has issued interpretive guidance for publicly held companies on disclosing to investors any serious risks that climate change poses to their operations (Securities and Exchange Commission, 2010).

Increasingly, companies strive to achieve ambitious environmental sustainability goals such as significant reductions in their energy use and greenhouse gas (GHG) emissions (Schendler, 2009; Unruh and Ettenson, 2010). Leaders across various industries have proposed making extreme cuts in carbon dioxide (CO_2) emissions, solid waste output, and energy and water usage. For example, Johnson Controls aims to reduce GHG emissions intensity by 30 percent over a ten-year period. The company's "Design for Sustainability" program monitors a product's environmental impact throughout the design, manufacturing, use, and disposal stages of its lifecycle (Johnson Controls, 2011). Bayer is setting similar goals, aiming for a 35 percent GHG intensity reduction over 15 years (Bayer AG, 2011). In manufacturing industries, where emissions are more difficult to reduce, the multinational building materials company Lafarge aims to cut its CO_2 emissions per ton of cement by 20 percent (Lafarge, 2008). Retail stores have very little control over the manufacturing of their merchandise, but Kohl's is now purchasing enough green power to cover 100 percent of its electricity use (King, 2011).

The development of environmental sustainability goals represents an important first step toward meeting the climate change challenge. By acting as a compass in determining the direction of work-related efforts, goals provide the means through which values lead to action (Lantham and Pinder, 2005). Although sustainability goals are very important (Milton and Stoner, 2008), converting intention into action requires more than motivation alone.

One factor that is likely to be central in moving firms from thought to action is how well a firm manages its knowledge resources. Knowledge management is increasingly important for organizations, which cannot survive in the competitive global economy without creating and transferring knowledge (Andrew, 2010). Scholars have long recognized that innovations arise through a process of recombining and extending existing knowledge (Schumpeter, 1934), which highlights the importance of knowledge transfer to knowledge-creation processes. In fact, organizations skilled at knowledge transfer have been found to be more competitive, more likely to survive, and more productive than their less-skilled counterparts (Argote, 1999). Lafarge, for example, benefited when one of its facilities in Malaysia adopted an efficient production routine that one of its facilities in Turkey had developed (Perrin, Vidal, and McGill, 2006). Such instances of knowledge transfer, which occur when one organizational unit benefits from the experience of another (Argote and Ingram, 2000), not only contribute to organizational effectiveness, but also supply tools firms need to tackle the risks and capitalize on the opportunities that climate change creates.

Despite the clear value of knowledge transfer, organizations sometimes fail to take advantage of their knowledge resources. There are many industrial cases of efficient production routines, best practices, and innovative ideas that have remained obscured in their unit of origin because potential recipients fail to look outside their own units for new ideas. Katz and Allen (1982) refer to this occurrence as the "not invented here" syndrome. An example of ineffective knowledge transfer can be seen in plants at Kodak failing to implement a practice for creating an important component that was developed in the firm's research laboratory (Leonard-Barton, 1988). An example of ineffective knowledge creation can be seen in the costly delays Airbus encountered when its German and French design units failed to share knowledge critical for designing its A380 super jumbo (Clark, 2007). The Airbus chief executive Louis Gallois blamed national pride and banned national symbols from appearing on presentation materials stating, "when you have a flag you have always an issue of national identity" (Clark, 2007, p. 2). These anecdotes are consistent with a systematic study of over 100 attempts to transfer best practices across units of multinational organizations, which resulted in the study's author lamenting the gap that often exists between what is known in an organization and what is put to use (Szulanski, 1996). The frequency of this situation raises the question of what can be done to facilitate knowledge transfer because the climate change challenge is too great for organizations to reinvent the wheel and miss opportunities to develop innovations.

This chapter argues that the kind of knowledge transfer needed to meet the climate change challenge is facilitated by a special kind of relationship among organizational units. The defining characteristic of this relationship is that employees in the units share a psychological sense of belonging to an overarching unit, which is termed "superordinate social identity." In accordance with social psychological scholarship on social identity (e.g., Brewer, 1979; Gaertner and Dovidio, 2000), this conceptualization recognizes that this collective identity arises from employees' valuing that they belong to a superordinate group to such an extent that they draw part of their own identity from this group membership. As such, superordinate identity is a psychological state that results when employees from multiple organizational units feel a sense of belonging to, or identification with, an overarching organizational entity (e.g., transnational team, department, division, or entire organization).

Louis Gallois, the Airbus chief executive who banned national symbols, seemed to be striving for such a relationship between the firms' European units when he stated "I want, when I am in Hamburg, to feel at home like in Toulouse" (Clark, 2007, p. 2). When employees' in-group expands in such a way, so too does the boundary around the concept of "here," with the "not-invented-here roadblock" becoming the "invented-here highway." In this manner, superordinate identity is likely to promote the kind of knowledge transfer and creation needed for organizations aiming to take actions to deal with climate change.

This chapter is organized as follows: First, it theorizes about the role superordinate identity might play in managing the knowledge resources key to meeting the climate change challenge. In so doing, it summarizes evidence that superordinate identity facilitates knowledge transfer and creation. Next, it explores two conditions under which superordinate identity is especially likely to facilitate a firm's ability to capitalize on its knowledge resources for sustainability. The chapter concludes with an exploration of steps that managers can take to promote the development of strong superordinate identities within their organizations.

Superordinate Identity and the Management of Knowledge Resources

This first section draws on published scholarship to describe *why* and *how* superordinate identity is likely to facilitate the kind of knowledge transfer and creation that is needed to meet the climate change challenge. Social psychologists and organizational scholars have developed a considerable body of theories and evidence about the importance of social identity in organizational settings (for a review, see Ellemers, Gilder, and Haslam, 2004). In their seminal work on the topic Tajfel and Turner (1979) advanced the concept of social identity, a part of a people's identity derived from belonging to important social groups, and described its effects on intergroup attitudes and behaviors. Principal among these effects are more favorable attitudes toward, and better treatment of, members of one's own group than members of other groups. Brewer (1979), for example, found that people tend to perceive members of their own groups more positively than members of other groups (e.g., as more valuable, trustworthy, honest, and loyal). Numerous studies have corroborated evidence of such in-group favoritism (Dasgupta, 2004). Behavioral manifestations, for example, include, but are not limited to, in-group supporting cooperative behaviors and resource allocations (Brewer, 1979; Kramer and Brewer, 1984; Tyler and Blader, 2001).

The idea that a superordinate identity could facilitate more harmonious intergroup relations arose from the observation that in-group favoritism does not have to be limited to subgroups (e.g., work teams); identification with a superordinate group (e.g., corporate division) expands favoritism to include members of the common, overarching group (Kramer; 1991; Gaertner and Dovidio, 2000). In a series of behavioral experiments, social psychologists amassed considerable evidence that people evaluate members of another subgroup more favorably after the subgroups are merged into one superordinate group (e.g., Dovidio, Gaertner, Validzic, et al., 1997; Gaertner, Mann, Murrell, et al., 1989). The application of this work to organizational settings extends well beyond the relatively rare situations of group and corporate mergers. Not only is it possible to create a higher-order superordinate identity that includes lower-order constituent groups, it is increasingly common in organizational settings. Scholars have documented numerous cases of employees developing a psychological sense of belonging to superordinate groups (e.g., transnational teams, cross-functional teams, organizations), while still remaining members of distinct organizational units separated by geography (Dukerich, Golden, and Shortell, 2002; Hinds and Mortensen, 2005), disciplinary expertise (Dukerich, Golden, and Shortell, 2002; Van Der Gert and Bunderson, 2005), and corporate function (Sethi, 2000).

Building on the previously described social psychological and organizational research, one of the present authors and her colleagues developed and tested a theory of superordinate identity and knowledge transfer (Argote and Kane, 2009; Kane, Argote, and Levine, 2005; Kane, 2010). This chapter applies and extends the above theory to the domain of sustainability initiatives. In particular, superordinate identity is posited to facilitate knowledge transfer relevant to those initiatives primarily by increasing the quality of attention that members of one group allocate to another group's knowledge. Specifically, stronger identification with a superordinate group motivates knowledge consideration, or focusing attention on determining the value of another's knowledge (Kane, 2010).

Such consideration may have averted the previously mentioned examples of ineffective knowledge transfer at Kodak and Airbus. Had the Kodak plants, for example, allocated sufficient attention to determining the value of the practice developed in their own research laboratory, they would have likely conducted their own pilot test and learned that the practice cut processing time from weeks to days. Such quality of attention is especially important for sustainability initiatives because applying the lessons learned in one context to another is rarely a simple matter of copying exactly or choosing between clear, binary choices (Milton and Stoner, 2008). Instead, it may involve exploring a promising scientific innovation, such as an alternative fuel that, if fully developed, could yield considerable payoffs both environmentally and economically.

How and *why* superordinate identity might motivate groups to consider another's knowledge was explored in Kane (2010). Her reasoning suggests that knowledge consideration is a potentially risky, exploratory process because it involves allocating limited cognitive resources to another group's knowledge even though it may be of little relevance or value to one's own group. Nonetheless, such exploratory processes are strategically necessary because they are critical for long-term adaptation and survival (March, 1991). The presence of a strong superordinate identity is likely to increase the perceived benefits of engaging in knowledge consideration for two reasons. First, due to in-group favoritism, recipient groups are likely to have a positive expectation regarding the value of knowledge from sources within their superordinate group. Second, recipient groups may strive to discover the value of knowledge from in-group sources because doing so contributes to a more positive social identity. Knowledge is a valuable resource, and groups with more valuable resources are generally sources of positive social identity. In other words, knowledge consideration offers the promise of uncovering valuable knowledge from a group within one's superordinate identity and thereby provides an avenue for fulfilling a desire for positive identity.

A central assumption of this theory is that knowledge transfer is a mindful process in which employees perceive some value in knowledge before they implement, adapt, or build on it. Value can be assessed along a number of dimensions including efficiency, aesthetics, economic impact, and environmental impact. The extent to which organizational members agree on the relative importance of dimensions should lead to less conflict in determining the value of knowledge. The presence of agreed upon dimensions, however, does not suggest that such value recognition will be straightforward or objective because, prior to adoption, the usefulness of knowledge can only be estimated (Menon and Blount, 2003). One factor likely to increase value recognition, which will be discussed in greater detail below, is the extent to which the merits of knowledge are apparent, termed "knowledge demonstrability" (Kane, 2010). Another factor is the extent to which units allocate attention to consider knowledge thoroughly enough to recognize its merits. Paying greater attention to another's knowledge (i.e., knowledge consideration) should improve value recognition, especially when knowledge is low in demonstrability with concealed merits, which is often the case for new, complex initiatives.

Evidence of Superordinate Identity and Knowledge Management

The theoretical perspective above is supported by research evidence indicating that superordinate identity facilitates the kind of knowledge transfer and creation that is needed to meet the climate change challenge (Argote and

Kane, 2009; Kane, Argote, and Levine, 2005; Kane, 2010). Superordinate identity appears to increase knowledge consideration, especially in cases when the value of that knowledge is not readily perceived (Kane, 2010). Examples of this phenomenon can be seen across industries, in cases where superordinate identity aided knowledge transfer and alternately where absence of a superordinate identity hindered knowledge transfer. Environmental sustainability initiatives in particular can be aided by superordinate identity because the complex and unfamiliar technologies sometimes used in these initiatives can increase the difficulty of perceiving the inherent value in such knowledge.

A first step in establishing that superordinate identity influences knowledge transfer through knowledge consideration is finding stronger evidence of the causal relationship when knowledge quality is higher rather than lower. Indeed, when Kane, Argote, and Levine (2005) conducted the first behavioral experiment testing the theory, they found such evidence. The researchers were able to establish causal relationships by systematically varying predictors (e.g., superordinate identity, knowledge characteristics) and observing their influence on outcomes (e.g., knowledge transfer). Superordinate identity was found to influence knowledge transfer more positively when knowledge was superior, rather than inferior, to recipients' existing knowledge.

A subsequent experimental study focused directly on whether superordinate identity influences knowledge transfer through the process of knowledge consideration (Kane, 2010). The study investigated whether knowledge consideration would improve value recognition, especially when knowledge is low in demonstrability with concealed merits. Superordinate identity was found to influence knowledge transfer more positively when knowledge was less demonstrable with concealed merits than when knowledge was high in demonstrability with apparent merits.

Kane (2010) provided further evidence of the underlying role of knowledge consideration, which can be measured using techniques like counting words in verbal protocols (Ericsson and Simon, 1980). In particular, examining group conversations provides a natural window into group information processing (Weingart, 1997). Consequently, a quantitative measure of knowledge consideration was obtained by systematically coding each line of group conversation (for details, see Kane, 2010). As expected, superordinate identity positively influenced knowledge consideration. Recipient groups focused significantly more of their limited attention on determining the value of another group's knowledge when both groups shared a superordinate identity than when both did not. In particular, compared to those without superordinate identity, those with such an identity engaged in three times as much knowledge consideration, which was observable from members' requests for and provision of information on routine characteristics, evaluations of the routine, and discussion of the merits of its adoption. Furthermore, mediation analyses revealed that amount of knowledge consideration accounted for the effect of the randomly assigned condition (i.e., superordinate identity, no superordinate identity) on knowledge transfer, particularly when knowledge was low in demonstrability with concealed merits.

Evidence from industrial settings is also consistent with the theory that superordinate identity leads to greater knowledge transfer and creation. Organizational researchers have documented cases of effective knowledge transfer in a wide range of industries including agriculture (Ingram and Simons, 2002), automotive (Epple, Argote, and Devadas, 1991), fast food (Darr, Argote, and Epple, 1995), hospitality (Ingram and Baum, 1997), information technologies (Cummings, 2004), and pharmaceuticals (Bresman, 2010). Looking across scholarly evidence of effective knowledge transfer, one observes that units benefit more from the experience of other units when they belong to a superordinate group such as a franchise (Darr, Argote, and Epple, 1995), chain (Ingram and Baum, 1997), or federation (Ingram and Simons, 2002). For example, pizza restaurants benefited from adopting an innovative process for assembling pizzas that had been developed at another unit within the same franchise (Darr, Argote, and Epple, 1995). Although supportive of the theory, none of these studies included measures of superordinate identity, meaning that greater knowledge transfer within superordinate groups could be due to other features that also occur when units are part of a superordinate group.

One field study has examined the association between superordinate identity and the exploration of a promising early-stage scientific innovation that if developed could help firms not only mitigate the risks of climate change, but also capitalize on its opportunities (Argote and Kane, 2009). As part of a larger study exploring the potential introduction of cellulosic ethanol as an alternative fuel for the US motor vehicle fleet, Argote and Kane (2009) obtained measures of strength of superordinate identity from a sample of ten employees at automobile and energy companies. Consistent with their theory of superordinate identity and knowledge transfer and creation, they expected that firms in these industries whose members viewed themselves as belonging more to a superordinate group (of producers irrespective of fuel type) rather than to a subordinate group (of petroleum-based fuel producers) would report a higher likelihood of introducing a nonpetroleum fuel like cellulosic ethanol.

Superordinate identity was measured by asking automobile employees to indicate the extent to which they felt that they belonged to a superordinate group of producers of cars irrespective of fuel type or a subgroup of producers of gasoline-powered cars. Likewise, energy employees indicated the extent to which they felt that they belonged to a superordinate group of suppliers of energy irrespective of source or a subgroup of suppliers of petroleum-based energy. An examination of how these responses related to employees' perceptions that their firms should introduce cellulosic ethanol in the next three years revealed a strong and significant correlation. A stronger superordinate identity was associated with a greater openness to adopting, adapting, and building on an innovation developed elsewhere that also had the potential to change defining characteristics of the firms themselves (e.g., energy firms often refer to themselves as oil and gas firms).

When Superordinate Identity is Especially Likely to Assist Firms in Meeting the Climate Change Challenge

The extent to which knowledge is demonstrable with apparent merits and the extent to which a firm has embraced sustainability goals are two factors that are likely to influence how much superordinate identity helps firms capitalize on the opportunities and mitigate the risks of climate change. Strong superordinate identity can be especially valuable for meeting the climate change challenge. The members of an organizational unit such as a team, plant, or division are unlikely to implement, adapt, or build on a sustainability initiative developed at another of the firm's units unless those members recognize value in the initiative. Knowledge transfer thus requires that potential recipients perceive value in adopting another's idea, routine, or practice. The kind of thorough consideration that arises from a superordinate identity is likely to be more important to value recognition in some cases than in others. Two conditions that are likely to influence whether the knowledge consideration is needed for people to perceive value in others' sustainability-related knowledge are the extent to which knowledge is demonstrable with apparent merits and the extent to which the firm has embraced sustainability goals.

Knowledge Demonstrability

People need to perceive value in another's routine, practice, or innovation before they will implement, adapt, or build on it. What then contributes to people perceiving value in other's sustainability-related knowledge? Clearly the initiative needs to have some intrinsic value, but objective characteristics are rarely sufficient to create uniform perceptions of value. There is one relatively rare condition under which people are likely to perceive the value of knowledge, and that is when knowledge is highly demonstrable with apparent merits. Knowledge that is high in demonstrability has merits that potential recipients can plainly see without allocating attention resources to determine its value. By contrast, knowledge low in demonstrability has merits that recipients cannot see unless they allocate significant attention resources to determine its value. As previously discussed, theory and evidence suggest that superordinate identity increases knowledge consideration, which is the kind of attention recipients must allocate to recognize the value of knowledge with concealed merits. Consequently, superordinate identity facilitates the transfer of less demonstrable knowledge, but does little to aid the transfer of highly demonstrable knowledge.

Sustainability initiatives with apparent merits are likely to transfer within an organization regardless of the presence of a superordinate identity. As previously mentioned, one unit of the multinational building materials company Lafarge adopted an efficient production routine that was developed in another one of the firm's facilities half a world away (Perrin, Vidal, and McGill, 2006). If the routine had apparent merits (e.g., low implementation costs, clear payoffs), decision makers at the Malaysian plant may not have needed to share a superordinate identity with their colleagues who developed the routine in Turkey. The Malaysian plant's decision makers may not have needed to allocate much attention to determine the value of the routine, which diminished the importance of superordinate identity.

Routines and practices that reduce energy consumption have the potential to be high in demonstrability because their adoption probably has apparent economic and environmental benefits. For those benefits to be apparent, however, the upfront economic cost of implementation likely has to be low relative to the expected payoff. Adopting a new work routine or practice is more likely to have a lower upfront expenditure than adopting a new piece of equipment. Changing the way employees work so that many may work from home, for example, has the potential to reduce greenhouse gases significantly with little upfront investment. By contrast, exchanging a facility's light bulbs for more energy-efficient ones may involve a sizeable upfront investment. This required investment may explain one reason why the sustainability officer at Aspen Ski Company encountered such resistance to replacing inefficient incandescent light bulbs with compact fluorescent light bulbs (Schendler, 2009). Both teleworking and replacing light bulbs have intrinsic economic and environmental value, but what appears to differ between these practices is their demonstrability.

Sustainability initiatives with less apparent merits are unlikely to transfer within an organization without the presence of a superordinate identity. As previously mentioned, the Airbus design teams in Germany and France failed to transfer knowledge critical for developing the A380 jumbo-jet (Clark, 2007). Because the knowledge likely had fewer apparent merits (e.g., implementation costs, unclear payoff), the collaborators may have needed to share a superordinate identity to encourage them to allocate significant attention to determining the value of each other's ideas.

Many climate change initiatives will very likely be low in demonstrability, rendering superordinate identity especially important. As previously discussed, even routines and practices that reduce energy consumption with economic and environmental benefits will be low in demonstrability when their merits are concealed by impediments such as initial costs. Such costs are quite common as many practices involve the upfront purchase of technological equipment. As previously discussed, such costs may obscure the merits of practices as uncontroversial as replacing energy-inefficient light bulbs with efficient ones. More complex practices are likely to be even lower in demonstrability. For example, although replacing travel to internal meetings with teleconference meetings could significantly reduce GHG emissions and economic costs generated from airline travel (Kitou and Horvath, 2008; Toffel and Horvath, 2004), the merits of this practice may be concealed by factors including upfront equipment costs.

The radically innovative practices, routines, or technologies that may be critical for meeting the climate change challenge are likely to be even lower in demonstrability, thus making superordinate identity even more important. Examples of these types of knowledge include alternative fuels such as cellulosic ethanol that are still being developed and that would require a major overhaul of how work is done. As previously discussed, employees in energy and automobile firms with stronger superordinate identities were more receptive to exploring this kind of knowledge. This quality of attention is needed when decision makers are faced with meeting the full complexity of the climate change challenge. The radical innovations that are needed to achieve substantial reductions in GHG emissions will require significant changes in the way we conduct business both in terms of how we consume and produce energy. Such significant changes increase uncertainty and thus require thoughtful attention to the new ideas, routines, and practices generated by others. We cannot possibly meet this challenge without learning from one another's experiences, and superordinate identity is one factor likely to facilitate such learning.

Environmental Sustainability Goals and Employee Commitment

A key requirement for implementing, adapting, or building on another's routine, practice, or innovation is that people perceive value in that knowledge.

As discussed in the previous section, knowledge demonstrability influences how much superordinate identity matters in value recognition. The kind of knowledge consideration that arises from superordinate identity is especially important for recognizing the merits of the many sustainability-related initiatives that are low in demonstrability. Attributes of knowledge clearly have a role to play in this perception; but so too do attributes of the perceiver. Recipients must perceive value in the knowledge. One important recipient attribute likely to influence this perception is the extent that the recipient's actions are driven by goals that are aligned with the knowledge, while another is the extent that the recipient's goals are aligned with the changes brought about by adopting the knowledge. Superordinate identity's influence on knowledge consideration and, in turn, on the adoption of sustainability-related knowledge is likely to depend on the extent to which recipients are committed to environmental sustainability goals.

People will value an environmental sustainability initiative differently depending on the extent to which they are committed to environmental sustainability goals. Motivational scholars describe how goals provide a compass directing efforts in one direction rather than another (Lantham and Pinder, 2005). In cases of knowledge transfer and creation, recipients rely on their goals to define important attributes of a routine, practice, or innovation. For example, efficiency was the primary goal of participants in the behavioral experiments described earlier (Kane, Argote, and Levine, 2005; Kane, 2010). Participants who shared a superordinate identity with a knowledge source considered the source's production routine thoroughly enough to recognize if it was more efficient than their own, and when it was, they adopted it. This work suggests that people will be more likely to perceive value in a sustainability-related idea, routine, practice, or innovation when they are more strongly committed to environmental sustainability goals than when they are less strongly committed to such goals.

A field study of hundreds of episodes of environmental championing that occurred mainly in the United States manufacturing and electricity industries provides evidence supportive of this point (Andersson and Bateman, 2000). These researchers found that it was much more difficult to champion environmental initiatives at companies whose members were less strongly committed to environmental goals, that showed little concern for the environment, or that espoused sustainability values only for the purposes of compliance or publicity. An inspirational appeal for an environmental issue was more effective in firms with strong environmental goals. The presence of strong superordinate identity will be more helpful in moving firms from thought to action regarding climate change when members are more committed to environmental sustainability goals.

A first step in understanding the strength of an organization's commitment to environmental sustainability goals is seeing whether the firm has espoused environmental goals. One might then assess the motivational quality of these goals. Goal-setting research indicates that motivating goals are specific and challenging (Lantham and Pinder, 2005). With regard to challenging environmental goals, scholars have pointed out that challenging goals can be especially inspiring (Milton and Stoner, 2008).

A second step could be to determine how the organization views environmental management. Andersson and Bateman (2000) have discussed a transition from a more traditional to a more progressive type of environmental management. Whereas traditional organizations view nature as a collection of resources to be managed, progressive organizations elevate nature to a higher level and in doing so create more challenging environmental goals. Organizations in the former paradigm might aim to include environmental criteria on the balance sheet and engage in recycling and waste management. Organizations in the latter paradigm might go further by aiming for sustainability innovation and environmental performance. These paradigms represent parts of a continuum, rather than discrete states of business practice. Many environmentally responsible organizations are moving in the direction of the progressive management model, which is likely to translate into ambitious environmental goals with greater motivating potential.

Next, one might evaluate the extent to which these goals directed action within the firm. Sustainability reports that highlight progress toward meeting such goals provide some data. Another indicator that the firm is committed to these goals is the presence of a chief environmental sustainability officer or unit within the firm. The presence of such a formally defined position, however, does not necessarily reflect decision-making power or strong goal commitment (Schendler, 2009). For that reason, some have argued that evidence of stronger environmental goal commitment can be seen in a sustainability officer position that is elevated to the level of direct report to the chief executive (Schendler, 2009). Indeed, the economic and social pillars of the Triple Bottom Line (Savitz, 2006) often have a seat in the firm's top management, embodied in the chief financial officer and the director of human resources. Related to the Triple Bottom Line, members of a firm who derive economic and social value from the actions that lead to environmental goal achievement should be more committed to environmental goals.

To conclude, the second of the two conditions under which superordinate identity is more or less likely to aid knowledge transfer and creation is the extent to which members of an organization are committed to sustainability goals. Goals provide a metric along which people evaluate ideas, practices, and initiatives. Consequently, employees who are more committed to environmental goals are more likely to benefit from the kind of knowledge consideration that arises from such an identity than employees who are less committed to such goals.

Developing Strong Superordinate Identities within Organizations

Social identity is a part of people's identity that they derive from important groups (Tajfel and Turner, 1979). We thus need to reflect on what makes identification with a particular group likely, as people simultaneously belong to multiple social groups. Scholarship indicates that people identify with a group when it is subjectively important or when membership in the group matches situational features (Haslam, 2004). Consequently, managers can support the development of strong superordinate identities by highlighting the superordinate group and by helping employees to see its value.

Highlighting a superordinate group, such as a geographically distributed or cross-functional team, department, division, or organization involves bringing the superordinate group to the foreground and pushing subgroups to the background. Powerful approaches for doing so include superordinate-level performance evaluations matched with monetary incentives and rewards (Argote and Kane, 2009; Kane, 2010; Kane, Argote, and Levine, 2005). By banning national flags on internal presentation materials, the Airbus CEO was attempting to push subgroups out of the foreground and into the background. Some modern management practices such as dynamic seating arrangements (e.g., hot desking) and telework pull individuals away from close-knit working relationships, which consequently highlights a higher-order group as employees find themselves attaching greater importance to their organizational membership (Millward, Haslam, and Postmes, 2007; Wiesenfeld, Raghuram, and Garud, 2001).

Research on employee selection indicates that employees will be more attracted to and likely to remain in organizations whose cultures match their personality (Schneider, Goldstein, and Smith, 1995). One way to enhance superordinate identity may be to select employees whose personal attributes meet those of a superordinate group. Thus, an important step would be to determine the defining characteristics of the unit that are likely to be appreciated by members (Dukerich, Golden, and Shortell, 2002). Superordinate groups whose defining characteristics are related to sustainability goals should be especially beneficial. To the extent that an organization has inconsistent characteristics across its lower-order units and managers are attempting to reinforce higher-order characteristics, it may be beneficial to conduct employee recruiting at a higher, rather than lower, level. Research also indicates that on average people identify more strongly with organizations with positive external images or reputations (e.g., Dukerich, Golden, and Shortell, 2002). This work suggests that marketing and public relations that tout the achievements of a superordinate group could also be effective.

Conclusion

Humans are part of the planet's naturally occurring ecosystem (Socolow, 1996). To appreciate this ecosystem fully, human endeavors must be considered natural phenomena. Manufacturing, resource harvesting, and the activities of corporations are a natural result of human society and are not separate from the natural environment. It is necessary for both individuals and corporations to broaden their views of human interaction with the environment to understand and ultimately solve today's environmental problems. As the climate change challenge becomes more pressing, corporate goals to combat environmental problems will become more common and more ambitious. Wide-reaching, strategic goals are inherently more complex because they require a multipronged, coordinated approach, incorporating all members of the organization. Integrating such a plan across multiple disciplines and departments will have a greater cumulative effect than any one action that is more narrowly focused.

Increasing participation in an environmental initiative requires that organizational members understand the inherent value in knowledge in order to adopt it, and also see their role within a larger group working toward a common goal. Superordinate identity is helpful in increasing knowledge consideration, which is especially important in situations where knowledge demonstrability is low. Low knowledge demonstrability often occurs in sustainability initiatives when the situation is complex and long-term value can be masked by conditions like upfront financial cost and resistance to changes in routine. Similarly, the more ambitious a goal, the more knowledge transfer and creation will be required from throughout an organization to accomplish it. Superordinate identity should be especially important in such cases because it increases the kind of knowledge consideration that is needed for employees to recognize the value in other's sustainability-related ideas.

Perceptions and goals of the environmental movement have been shifting from reactive to proactive, from isolated to holistic. While this transition is a crucial first step, a change in theory is moot without a change in practice.

Fostering a superordinate identity within an organization will increase the quality of attention given to others' ideas and thereby aid knowledge transfer and communication within the larger group, while also enabling an improved understanding of how to work together to achieve common, ambitious goals. To bring about the radical change necessary to combat climate change effectively, people across organizations and nations may need to develop a sense of a superordinate identity that is tied to a sense of responsibility to a greater purpose. Along these lines, researchers found in a behavioral experiment that people from around the world contributed more of their limited resources to a collective world account when they had a stronger rather than weaker sense of global identity (Buchan, Brewer, Grimalda, et al., 2011). This type of unified effort may be the most effective method of creating a positive impact in the area of responsible resource use.

References

Andersson, L. M., and T. S. Bateman. 2000. Individual environmental initiative: Championing natural environmental issues in U.S. business organizations. *Academy of Management Journal* 43: 584–570.

Andrew, J. P. 2010, April 15. What executives make of innovation. *Bloomberg Business Week*. http://www.businessweek.com (Accessed July 31, 2011).

Argote, L. 1999. *Organizational learning: Creating, retaining and transferring knowledge*. Norwell, MA: Kluwer Academic Publishers.

Argote, L., and A. A. Kane. 2009. Superordinate identity and knowledge creation and transfer in organizations. In *Knowledge Governance*, ed. N. J. Foss and S. Michailova (pp. 166–190). Oxford: Oxford University Press.

Argote, L., and P. Ingram. 2000. Knowledge transfer: A basis for competitive advantage in firms. *Organizational Behavior and Human Decision Processes* 82: 150–169.

Bayer AG. 2011. Bayer policy on climate change. *Bayer Climate Program*. http://www.climate.bayer.com (Accessed July 31, 2011).

Baum, J. A. C., and P. Ingram. 1998. Survival-enhancing learning in the Manhattan hotel industry, 1898–1980. *Management Science* 44: 996–1016.

Bresman, H. 2010. External learning activities and team performance: A multimethod field study. *Organization Science* 21: 81–96.

Brewer, M. B. 1979. In-group bias in the minimal intergroup situation: A cognitive motivational analysis. *Psychological Bulletin* 86: 237–243.

Buchan, N. R., M. B. Brewer, G. Grimalda, R. K. Wilson, E. Fatas, and M. Foddy. 2011. Global social identity and global cooperation. *Psychological Science*, ePub ahead of print May 17, doi:10.1177/0956797611409590.

Clark, N. 2007, May 18. Turnaround effort is challenging at Airbus, a stew of European cultures. *The New York Times*. http://nytimes.com (Accessed July 31, 2011).

Cummings, J. N. 2004. Work groups, structural diversity, and knowledge sharing in a global organization. *Management Science* 50: 352–364.

Darr, E. D., L. Argote, and D. Epple. 1995. The acquisition, transfer, and depreciation of knowledge in service organizations: Productivity in franchises. *Management Science* 41: 1750–1762.

Dasgupta, N. 2004. Implicit group favoritism, outgroup favoritism, and their behavioral manifestations. *Social Justice Research* 17: 143–170.

Dovidio, J. F., S. L. Gaertner, A. Validzic, K. Matoka, B. Johnson, and S. Frazier. 1997. Extending the benefits of recategorization: Evaluations, self-disclosure, and helping. *Journal of Experimental Social Psychology* 33: 401–420.

Dukerich, J. M., B. R. Golden, and S. M. Shortell. 2002. Beauty is in the eye of the beholder: The impact of organizational identification, identity, and image on the cooperative behaviors of physicians. *Administrative Science Quarterly* 47: 507–533.

Ellemers, N., D. De Gilder, and S. A. Haslam. 2004. Motivating individuals and groups at work: A social identity perspective on leadership and group performance. *Academy of Management Review* 28: 459–478.

Environmental Information Administration. 2011. *Annual Energy Outlook*. http://www.eia.gov (Accessed July 31, 2011).

Epple, D. L., L. Argote, and R. Devadas. 1991. Organizational learning curves: A method for investigating intra-plant transfer of knowledge acquired through learning by doing. *Organizational Science* 2: 58–70.

Ericsson, K. A., and H. A. Simon. 1980. Verbal reports as data. *Psychological Review* 87(3): 215–251.

Gaertner, S. L., and J. F. Dovidio. 2000. *Reducing intergroup bias: The common in-group identity model*. Philadelphia: Psychology Press.

Gaertner, S. L., J. Mann, A. Murrell, and J. F. Dovidio. 1989. Reducing intergroup bias: The benefits of recategorization. *Journal of Personality and Social Psychology* 57: 239–249.

Haslam, S. A. 2004. *Psychology in organizations: The social identity approach* (2nd ed.). London: SAGE.

Hinds, P. J., and M. Mortensen. 2005. Understanding conflict in geographically distributed teams: The moderating effects of shared identity, shared context, and spontaneous communication. *Organization Science* 16, 290–307.

Ingram, P., and J. A. C. Baum. 1997. Chain affiliation and the failure of Manhattan hotels, 1898–1980. *Administrative Science Quarterly* 42: 68–102.

Ingram, P., and T. Simons. 2002. The transfer of experience in groups of organizations: Implications for performance and competition. *Management Science* 48(12): 1517–1533.

Johnson Controls Inc. 2011. Environmental Scorecard. http://www.johnsoncontrols.com (Accessed July 31, 2011).

Kane, A. A. 2010. Unlocking knowledge transfer potential: Knowledge demonstrability and superordinate social identity. *Organization Science* 21: 643–660.

Kane, A. A., L. Argote, and J. M. Levine. 2005. Knowledge transfer between groups via personnel rotation: Effects of social identity and knowledge quality. *Organizational Behavior and Human Decision Processes* 96: 56–71.

Katz, R., and T. J. Allen. 1982. Investigating the Not Invented Here (NIH) syndrome: A look at the performance, tenure, and communication patterns of 50 R & D project groups. *R & D Management* 12: 7–19.

King, B. 2011, May 25. Kohl's achieves net zero emissions. *Sustainable Business Weekly*. http://www.sustainablelifemedia.com (Accessed July 31, 2011).

Kitou, E., and A. Horvath. 2008. External air pollution costs of telework. *International Journal of Life Cycle Assessment* 13: 155–165.

Kramer, R. M. 1991. Intergroup relations and organizational dilemmas: The role of the categorization processes. In *Research in Organizational Behavior*, ed. L. L. Cummings and B. M. Staw (Vol. 13, pp. 191–228). Greenwich, CT: JAI Press.

Kramer, R. M., and M. B. Brewer. 1984. Effects of group identity on resource use in a simulated commons dilemma. *Journal of Personality and Social Psychology* 46(5): 1044–1057.

Lafarge. 2008. *Sustainability ambitions 2012*. http://www.lafarge.com (Accessed July 31, 2011).

Lantham, G. P., and C. C. Pinder. 2005. Work motivation theory and research at the dawn of the twenty-first century. *Annual Review of Psychology* 56: 485–516.

Leonard-Barton, D. 1988. Implementation as mutual adaptation of technology and organization. *Research Policy* 17: 251–267.

March, J. G. 1991. Exploration and exploitation in organizational learning. *Organization Science* 2(1): 71–87.

Menon, T., and S. Blount. 2003. The messenger bias: A relational model of knowledge valuation. In *Research in Organizational Behavior*, ed. B. M. Staw (Vol. 25, pp. 137–186). Greenwich, CT: JAI Press.

Mickley, L. J., D. J. Jacob, and D. B. Field. 2004. Effects of future climate change on regional air pollution episodes in the United States. *Geophysical Research Letters* 31.

Millward, L., S. A. Haslam, and T. Postmes. 2007. Putting employees in their place: The impact of hot desking on organizational and team identification. *Organization Science* 18: 547–559.

Milton, L. P., and J. A. F. Stoner. 2008. Toward environmental sustainability: Developing thinking and acting capacity within the oil and gas industry. In *Innovative Approaches to Global Sustainability*, ed. C. Wankel and J. A. F. Stoner (pp. 123–158). New York: Palgrave Macmillan.

Perrin, A., P. Vidal, and J. McGill. 2006. Valuing knowledge sharing at Lafarge. *Knowledge and Process Management* 13: 26–34.

Savitz, A. W. 2006. *The triple bottom line*. San Francisco: Jossey-Bass.

Schendler, A. 2009. *Getting green done: Hard truths from the front lines of the sustainability revolution*. New York: Public Affairs.

Schneider, B., H. W. Goldstein, and D. B. Smith. 1995. The ASA framework: An update. *Personnel Psychology* 48: 747–773.

Schumpeter, J. A. 1934. *The theory of economic development: An inquiry into profits, capital, credit, interest, and the business cycle* (R. Opie, Trans.). New Brunswick, NJ: Transaction Books.

Securities and Exchange Commission. 2010. *SEC issues interpretive guidance on disclosure related to business or legal developments regarding climate change* [Press release]. http://www.sec.gov/news/press/2010/2010-15.htm (Accessed July 31, 2011).

Sethi, R. 2000. Superordinate identity in cross-functional product development teams: Its antecedents and effect on new product performance. *Journal of the Academy of Marketing Science* 28: 330–344.

Socolow, Robert H. 1996. *Industrial ecology and global change.* Cambridge University Press.

Solomon, S., G.-K. Plattner, R. Knutti, and P. Friedlingstein. 2009. Irreversible climate change due to carbon emissions. *Proceedings of the National Academy of Sciences of the United States of America* 106: 1704–1709.

Stern, N. H. 2007. *The economics of climate change: The Stern Review.* New York: Cambridge University Press.

Szulanski, G. 1996. Exploring internal stickiness: Impediments to the transfer of best practice within the firm. *Strategic Management Journal* 17: 27–43.

Tajfel, H., and J. C. Turner. 1979. An integrative theory of group conflict. In *The social psychology of intergroup relations,* ed. W. G. Austin and S. Worchel (pp. 33–47). Monterey, CA: Brooks/Cole.

Toffel, M. W., and A. Horvath. 2004. Environmental implications of wireless technologies: News delivery and business meetings. *Environmental Science & Technology* 38: 2961–2970.

Tyler, T. R., and S. L. Blader. 2001. Identity and cooperative behavior in groups. *Group Processes & Intergroup Relations* 4: 207–226.

Unruh, G., and R. Ettenson. 2010, November. Winning the green frenzy. *Harvard Business Review* 110–116.

Van Der Vegt, G. S., and J. S. Bunderson. 2005. Learning and performance in multidisciplinary teams: The importance of collective team identification. *Academy of Management Journal* 48(3): 532–547.

Weingart, L. R. 1997. How did they do that? The ways and means of studying group process. *Research in Organizational Behavior* 19: 189–239.

Wiesenfeld, B. M., S. Raghuram, and R. Garud. 2001. Organizational identification among virtual workers: The role of need for affiliation and perceived work-based social support. *Journal of Management* 27: 213–229.

CHAPTER 11

Final Thoughts

James A. F. Stoner and Charles Wankel

In the spirit of the Stockdale paradox (Collins, 2001), the authors in this volume have been willing to "face the brutal facts" of climate change but have "never lost faith" that businesses and governments and each one of us can, and, hopefully, will start to do the "right things." The "brutal facts" of that near 100 percent scientific consensus are that climate change is real, is already happening, and will get worse, and that much (or all) of it is caused by the way we produce and consume on this planet and the way we have chosen to, many believe, overpopulate it. These are the brutal facts that we all must face.

In this final chapter we have recognized the grounded optimism of the authors of the earlier chapters and their ability to see "Global sustainability risk management as a portal to opportunities." They have built on their own research and experiences and the research and analyses of others in accepting that virtually unanimous global scientific consensus about climate change. But we are sure we are joining them in believing that "doing the right things" involves every one of us, every business, and every government facing those brutal facts and taking bold, aggressive—yet compassionate, sometimes risky, and often experimental—actions to reduce the rate of climate change and to adapting to the negative impacts of those changes.

The suggestion that climate change also offers "a portal to opportunities" goes beyond the well argued, and we believe valid, point that there are ways for businesses to make money both by reducing climate change trajectories and by adjusting to those changes that have already occurred and will continue to occur. Our perception is that many, but not all, of the "right things" offer very real potential benefits. The right things done right can add to the quality of individual lives, the health of many new and old businesses, and improve the performance and contributions of many governmental activities, while also increasing governmental revenues and/or decreasing costs. And, for those "right things" that do not improve our lives in traditional terms, do not improve traditional measures of corporate profits, and/or do not increase government revenues or decrease government costs, we now argue that our lives can often be improved by "doing the right things" simply because they are the right things to do.

We join this book's authors in concluding that the dangers, risks, ambiguities, and uncertainties of climate change are very great indeed, but that the opportunities for the transformation of the ways our species inhabits *Eaarth* (McKibben, 2010) at present and the type of world we will leave for our children, grandchildren, the seventh generation, and beyond are inspiring and breathtaking—both the greatest challenge our species has ever faced and the "only game big enough for any of us to play."

References

Collins, J. C. 2001. *Good to great: Why some companies make the leap—and others don't.* New York: HarperBusiness.

McKibben, B. 2010. *Eaarth: Making a life on a tough new planet.* New York: Times Books.

Index